Cycles of Carbon - Nature's Balancing Act

Edited by Vinay Kumar

Published in London, United Kingdom

Cycles of Carbon - Nature's Balancing Act
http://dx.doi.org/10.5772/intechopen.1004692
Edited by Vinay Kumar

Contributors
Ganesh Thakur, Jerome Ntege, Martina Gaglioti, Nabulya Jesca, Peter Oseghale Ohue, Sabyasachi Prakash, Süleyman Akçay, Vihaan Rajput, Vinay Kumar, Xuejia Du

First published in London, United Kingdom, 2025 by IntechOpen
IntechOpen is the global imprint of INTECHOPEN LIMITED, registered in England and Wales, registration number: 11086078, 167-169 Great Portland Street, London, W1W 5PF, United Kingdom

For EU product safety concerns: IN TECH d.o.o., Prolaz Marije Krucifikse Kozulić 3, 51000 Rijeka, Croatia, info@intechopen.com or visit our website at intechopen.com.

British Library Cataloguing-in-Publication Data
A catalogue record for this book is available from the British Library

Cycles of Carbon - Nature's Balancing Act
Edited by Vinay Kumar
p. cm.

This title is part of the Earth Sciences Book Series, Volume 7
Series Editor: Maurizio Lazzari

Print ISBN 978-1-83634-179-6
Online ISBN 978-1-83634-178-9
eBook (PDF) ISBN 978-1-83634-180-2
ISSN 3049-8848

If disposing of this product, please recycle the paper responsibly.

IntechOpen Book Series

Earth Sciences

Volume 7

Aims and Scope of the Series

The world of Earth Sciences, considering the interactions within the geosphere and between the geosphere–biosphere, is a place in which a large number of scientists find and have found over time their own relationship or sector of application precisely because it is absolutely transversal to many disciplines and subdisciplines, which do not necessarily fall within the Geosciences. The objective of this book series is to welcome original scientific contributions both in consolidated contexts and in new frontiers of research, as well as review papers included in the various disciplines of Earth Sciences, but above all, those that show a modern and transversal vision of applications and impacts on the community in a particular historical context, which, following the COVID-19 pandemic, has shifted global attention to sectors that were previously more neglected. In particular, those of mining research and fossil and renewable energy sources, environmental geology and the sustainable use of natural resources and impacts on the built environment, land consumption, geoarchaeology, forensic geology, geotourism/geoheritage, georisks and climate and environmental changes, considered at different scales, up to new applications of geostatistical and geospatial analysis, GIS and artificial intelligence for the definition of forecasting models and scenarios in various sectors of basic and applied research.

Meet the Series Editor

Dr. Maurizio Lazzari has a Ph.D. in Earth Science and is a researcher at the Italian National Research Council, Institute of Cultural Heritage Sciences. Since 2001, Dr. Lazzari has been a Professor of Pedology at the University of Basilicata (Italy) and a geoarchaeologist at the University of Salento (Italy). His research activities are focused on natural and anthropic hazards and risk factors, aimed at safeguarding and conserving settlements and the historical–monumental heritage of the Mediterranean, with particular attention to landslide processes, susceptibility maps, monitoring, and modelling. Since 2004, he has been working as a scientific coordinator for several national research projects studying landslides and triggering factors, natural and anthropogenic risks, geological and geomorphological mapping, soil erosion, preservation of historical and archaeological sites, enhancement of degraded areas, geo-touristic use, and the protection of the landscapes. He is the author of about 150 scientific publications in national and international journals, monographs, book chapters, and conference proceedings concerning applied geology, geomorphology, dynamics of artificial reservoirs, soil erosion, landslides, geoarchaeology, hydrogeological instability, natural hazards, monitoring, cultural landscape, UNESCO Heritage, geoarchaeology, and geo-tourism.

Meet the Volume Editor

Dr. Vinay Kumar is an Assistant Professor of Meteorology at the University of the Incarnate Word, where he teaches weather-climate dynamics and researches atmospheric science with a focus on monsoonal droughts and climate modeling. As a prolific scholar, he has authored over 55 peer-reviewed articles, ten book chapters, and edited two books, with his work published in leading journals (AMS, AGU, Springer, Wiley) and featured as a guest editor for MDPI and JMSJ. Beyond academia, he is an avid sports enthusiast, playing cricket, tennis, and billiards. He finds balance in gardening and family time. Driven by a deep commitment to community impact, Dr. Kumar collaborates with students and colleagues to tackle global climate challenges through science and shared purpose.

Contents

Preface

Years ago, while analyzing a routine atmospheric dataset, I stumbled upon an anomaly, a tiny, persistent spike in CO_2 concentrations that shouldn't have been there. At that moment, a flicker of unease took root in me. It wasn't just a number but a fingerprint, a sign that our planet's ancient carbon rhythms were being rewritten. This book grew from that unease from my two-decade journey as a meteorologist chasing carbon's invisible footprints across skies, oceans, and policy papers.

We begin where my students always insist, we must: with why these matters. Chapter 1 isn't just about sustainability, it's about survival. I've stood in classrooms where teenagers gasp at graphs of hockey-stick curves and in boardrooms where executives glaze over at the same data. Bridging that gap demands stories, not just statistics. Here, we confront how industrial legacies became existential debt and why education is our best shovel to dig us out.

Money moves the world, and Chapter 2 exposes carbon finance's paradoxes. (I learned this the hard way during a frustrating 3 a.m. negotiation at COP27, watching delegates haggle over carbon credit while island nations drowned in real-time.) The numbers show promise, investment flows and market mechanisms are in place, but the ground reality remains unchanged. We're still treating the symptoms, not the disease.

Then, hope: Chapter 3 whisks us to Indonesia's mangroves, where I once waded knee-deep in brine alongside fishermen-turned-carbon-accountants. These 'blue forests' aren't just climate solutions; they're living libraries that teach us resilience. But as a typhoon survivor once yelled over crashing waves during my fieldwork: "You scientists love measuring, when will you fix?"

That question haunts Chapter 4's exploration of carbon storage. In Rajasthan's oilfields, engineers showed me how depleted wells could become ingenious yet bittersweet carbon vaults. Technology isn't a knight in shining armor; it's a rusty tool we're desperately reforging mid-battle.

Precision matters: Chapter 5's groundwater study near a Texas DAC site was born from coffee-stained well logs I pored over with a hydrologist who'd monitored that aquifer since the '70s. "Every ppm tells a story," she'd say. Our solutions must listen to those stories or risk creating new crises while solving old ones.

This book isn't a manifesto or an eulogy. It's a field notebook from humanity's messy, magnificent struggle to rebalance its relationship with carbon. It is one part lab coat and one part mud-stained boots, and it is entirely unfinished. The data matters, but so do the fishermen, the engineers, the students, and you holding these pages. Turn them with purpose.

I sincerely thank the IntechOpen publishing team, especially Ms. Laura Divic, Publishing Process Manager, for their invaluable guidance and support throughout this rewarding publication journey. Their professionalism and timely assistance made the entire process both smooth and enjoyable.

Vinay Kumar
Department of Atmospheric Science, Environmental Science and Physics,
University of the Incarnate Word,
San Antonio, USA

Section 1

The Problem and Global Systems

Chapter 1

Sustainability in the Context of Environmental Education

Süleyman Akçay

Abstract

The world today has entered a process that seems unlikely to end well for the environment, starting with the industrial revolution. During this process, the effects of excessive carbon dioxide and other greenhouse gases (such as methane) released into the atmosphere have manifested themselves as global warming and climate change. In addition, pesticide waste, water and air pollution, raw materials, and clean energy problems have also emerged locally. The most logical and holistic solution to this seems to be to switch to a sustainable development model. In this section, sustainability is discussed within the framework of environmental education.

Keywords: environmental education, sustainability, climate change, environmental literacy, education for sustainable development

1. Introduction

Energy is one of the most important needs of individuals in today's world. In addition, the economic growth and development of countries are compared with their energy consumption [1]. The increase in energy use has enabled economic growth to increase and needs to be met quickly. This has put the world in a vicious circle in terms of energy demand [2]. The energy source of developed and developing countries is predominantly based on fossil fuels [3–5]. This rate is 81.5%, according to 2023 data [6]. Consumption of fossil fuels increases the carbon dioxide rate in the atmosphere. The existing carbon dioxide rate in the atmosphere has shown dramatic increases since the Industrial Revolution. This has brought about many problems, especially global warming and climate change. Examples of these problems include a decrease in biodiversity, an increase in floods and hurricanes, drought, desertification, rising sea levels, and inability to access sufficient drinking water [7]. The increase in carbon dioxide in the atmosphere from fossil fuels also has an indirect effect. This effect leads to the melting of ice on the polar ice-covered lands (which have six methane deposits) and causes the release of additional methane gas into the atmosphere (**Figure 1**) [9].

Global warming caused by greenhouse gases such as carbon dioxide and methane also increases ocean water temperatures. This reduces the solubility of oxygen in ocean waters. Low oxygen and carbon dioxide produced by cellular respiration by consumers in the water cause this water to become acidic [7].

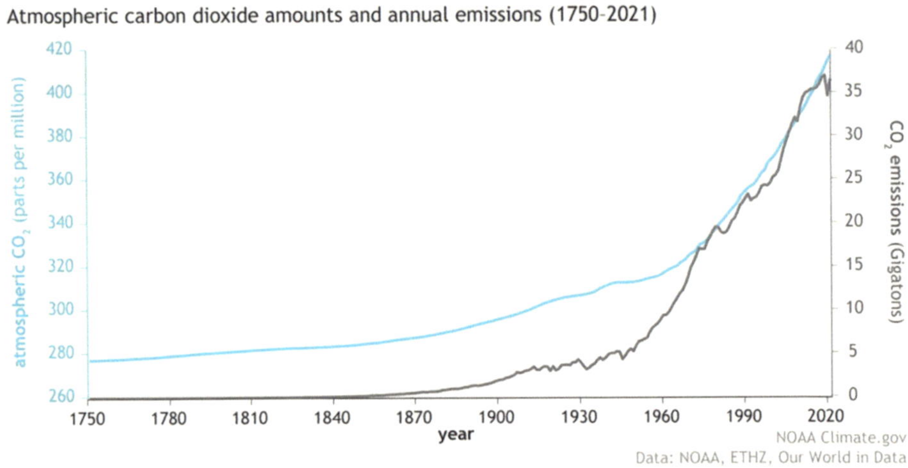

Figure 1.
Dramatic change in atmospheric carbon dioxide emissions after 1950 [8].

With the Industrial Revolution, the world entered a period in which natural resources were rapidly consumed without considering the future and the environment [10]. In addition, the Industrial Revolution somehow made people selfish, greedy, and far from the perception of social equality [11]. It can be said that there was more production and more consumption frenzy without considering the needs of other living things [12]. According to Albritton Jonsson [13], this period (Anthropocene epoch) in which humankind's impact on the Earth has reached its highest levels began with the Industrial Revolution and continues. According to him, the most distinctive feature of this era is global warming and climate change caused by excessive carbon dioxide released into the atmosphere (**Figure 2**).

Worster [15] also implies that these dramatic changes in the environment and the increase in carbon dioxide emissions will lead to the end of the world. In addition, a significant amount of energy is consumed in the production processes of the vehicles (e.g., cars and their parts, household appliances) that we use in daily life (at home or

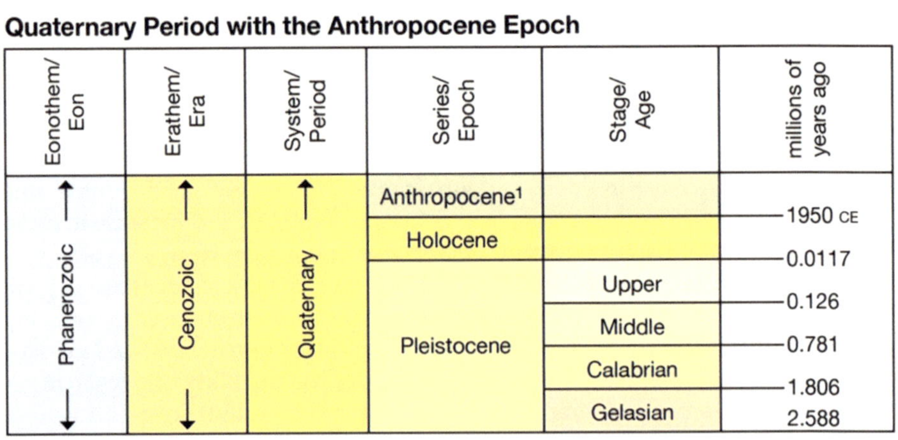

Figure 2.
Anthropocene epoch (the geological era in which the Earth is located since 1950) [14].

in industry). Again, energy is required in the process of obtaining clean water and transferring it to areas of use (home, industry, and agriculture). Another solution that can prevent this energy consumption is recycling [16]. Recycling is the process of making unused waste reusable or evaluating it as new and different materials [17]. The production of materials we use in every area of life from recyclable materials and thus recycling them for reuse is increasingly gaining importance in terms of energy and raw material sustainability [18]. Making recyclable materials that we frequently use in our daily lives, such as paper, glass, plastic, electronic waste, batteries, and organic waste, are concrete examples of recycling [19, 20]. Environmental pollution and the decrease in raw material resources, which are increasingly felt in the world, reveal the necessity of recycling [21, 22].

Until the 1980s, most governments viewed global environmental problems as unimportant issues in terms of both domestic and international politics. Later, the rise of environmental movements occurred, especially in industrialized countries. It can be said that the problems societies encountered with climate change were effective in this process. Dramatic decreases in fishing activities and public health problems caused by air and water pollution have given global environmental problems an important place in world politics. Today, environmental problems are important globally because they affect both the comfort of life of societies and other aspects of world politics, including security. Global concerns about the environment gained momentum with the announcement to the public of negative findings determined by scientists on the atmosphere, oceans, forests, land cover, and biodiversity [23]. The solution to the environmental threats exemplified above requires serious economy and labor. This cannot be solved by the unilateral activities of a single country. It requires a collective purpose and effort on a global scale [24–26]. The perception of this fact by society (led by science communication) and the activities of the intellectual segment of the society contribute to the positive evolution of the governments' global environmental policies [27–29]. Sustainable development and environmental protection efforts gained momentum with the work of the World Commission on Environment and Development [30] and the UN Conference on Environment and Development (UNCED) held in Rio de Janeiro, Brazil in June 1992. This was followed by a workshop held in Athens in 1995 to reconsider education for sustainable development [31]. The World Summit on Sustainable Development (WSSD) was held in Johannesburg in 2002. Finally, this was followed by the World Conference on Sustainable Development held in Rio de Janeiro in June 2012. All these events can be seen as important milestones in the advancement of sustainable development education with the aim of transitioning to quality environmental education [32]. In the 1990s, the concern brought about by environmental problems and the foresight of the possible problems they would bring gave birth to an education aimed at sustainable development. Although the Education for Sustainable Development (ESD) model of education aimed at sustainability has been adopted, even today, there are no guiding principles for the implementation of ESD goals in schools [33, 34].

2. Environmental literacy

The term environmental literacy was first used by Charles Roth in 1968 [35]. According to him, environmental literacy should be the primary goal of education rather than general literacy [36, 37]. Environmental literacy includes behaviors such as understanding the environment, having environmental knowledge and skills, and showing environmental attitudes. These behaviors enable individuals to establish a positive connection with their environment. They also ensure that the relationship between people and the areas in which they live is long-term and sustainable [35, 37].

The distinguished naturalist and nature writer Rachel Louise Carson's work "Silent Spring," published in 1962, triggered widespread environmental awareness and concern in societies, especially in the United States [35]. In this work, Carson questioned the indiscriminate and excessive use of dichlorodiphenyltrichloroethane (DDT) in the United States without knowing its possible effects [38]. With the impact of Silent Spring, society began to discuss the unquestioning commitment to development without regard for the environment and without sustainability and to express their uneasiness and doubts about the future [39]. In this social atmosphere, Roth [36] raised a question about the definition of an environmentally literate individual. Over time, this term has been used more frequently in the field of environmental education [35]. Today, environmental literacy has a meaning beyond being knowledgeable [40]. Environmentally literate individuals should be willing to protect the environment and be ready to take action when necessary [41].

3. Education for sustainable development

There are two dominant trends in environmental education in the world today. The first is Environmental Education (EE), and the second is education for sustainable development (ESD). ESD is also supported by the United Nations in order to ensure that countries achieve sustainable development. In contrast, EE was the first educational model with the aim of environmental protection. ESD was adopted in order to improve and make it more efficient [42]. In today's busy and complex life, many people often forget that they and all of humanity, of which they are integral, are inextricably linked to the biosphere. According to Vernadsky [43], in today's intense and complex life, many people often forget that all of humanity is inextricably linked with the biosphere. Also, man is traditionally and stereotypically called an individual who moves freely in the world and freely builds his own history. So far, no scientist has consciously taken into account the natural laws of the biosphere of the Earth, the only place where life can exist. Man cannot be completely or partially isolated from the biosphere. And these undeniable relationships are only beginning to become clear today, when compared with the past history of mankind. In reality, there is no living organism on Earth that lives in a completely independent state. All these organisms in the biosphere are inextricably and constantly connected with their environment, primarily through nutrition and respiration (cycles of matter and energy) (**Figure 3**) [43].

In this context, there are studies implying the importance of a better understanding of the matter and energy cycles in the biosphere for effective and behavioral environmental education [45, 46]. However, principle-based scientific reasoning is the basic requirement here [47]. In this chapter, it is argued that the matter and energy cycles in the biosphere can be better understood with examples of principle-based scientific reasoning and argumentation reinforcement in terms of the efficiency of sustainable environmental education. We also draw attention to the importance of comprehending the ontological and necessary interdependence of humans and other living and non-living entities. In short, there is an energy and matter cycle between living beings and inanimate beings. Explaining this in formal education processes with a holistic perspective makes it easier to understand the nature of the

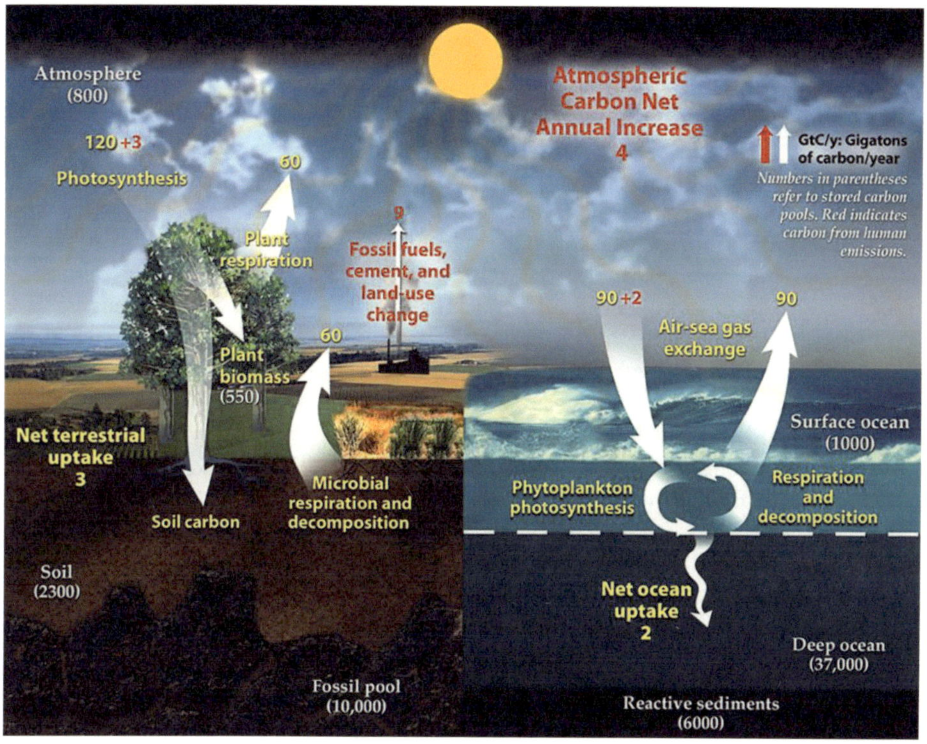

Figure 3.
Carbon cycle diagram with numbers (from Office of Biological and Environmental Research of the U.S. Department of Energy Office of Science) [44].

ecosystem's functioning [48]. In addition, these indispensable relationships are based on an ontological basis. For example, all living beings need the sun and producers for energy [49]. Only in this way can effective interventions be made in political decision-making processes with a collective consciousness. This can be a suggestion for formal education processes in general.

In other aspects (concerning the whole society), media and science communication are necessary. Here, scientists face the difficult but critical task of clearly communicating evidence-based information to the public and policymakers in order to positively and sustainably influence environmental policies [50, 51].

4. Discussion and conclusion

Environmental education is a complex but transdisciplinary educational strategy that aims for a general perspective, thought, and, ultimately, action that includes all disciplines [28]. Environmental education, which has evolved into sustainability education more predominantly today, aims to motivate all components of society to take action for sustainability now and in the future. In this sense, sustainability education for sustainable development has become the general motto in many countries [52]. Today, sustainability practices aim to protect natural resources and ecological balance for future generations. In addition, sustainability aims to develop actionable strategies rather than remaining theoretical [53–55].

Universities have an intellectual position in society. At the same time, political decision-makers usually receive a university education. In this sense, higher education institutions have an important role in spreading sustainable thinking within society [56]. Therefore, within the framework of the above-mentioned suggestions, innovations in higher education seem to be the most economical and reasonable way to achieve an effective result.

In addition, it is considered important to create eco-anxiety and ecological awareness among the whole society, especially in higher education. For this purpose, it may be useful to create and support discussion environments (concrete or abstract through media tools) that are scientific and far from hostile and comprehensive in their thinking [57]. In this context, most higher education students want to learn more about sustainable development. Similarly, they want to work in companies that actively support sustainable development in the future [58]. Despite this, very few higher education academics integrate sustainable development into their curricula in a comprehensive manner. In addition, the number of academics who carry out sustainability-related activities is quite low [59]. Despite their low knowledge of solid waste management, university students are willing to participate in recycling projects. In addition, sufficient economic incentives increase this willingness [60].

It is known that students and teachers at other formal education levels generally have positive environmental attitudes and high awareness of environmental problems [61, 62]. However, it seems that practical training of teachers is lacking in order to direct theory into practice (behavior). Here, there are positive linear relationships between the knowledge and attitudes of teachers and students [63]. It is known that environmental awareness is weak, especially in developing countries [63, 64]. Again, in relation to this, it has been determined that as the level of education increases in developing countries, people become more sensitive to the environment [65]. In terms of sustainability perception in the construction sector, countries also exhibit data that is directly proportional to their development. It is also claimed that social, economic, and environmental factors play a role in this [66]. Again, in developing countries, there is a linear positive relationship between environmental sustainability activities (e.g., reducing carbon dioxide emissions) and the rate of information and communication technology use in society [67]. In this sense, activities that will increase the use of communication technologies in societies, especially in developing countries, will be beneficial.

In recent years, many theoretical studies have been conducted on the competencies that a teacher will need in the education of sustainable development in the classroom [68–71]. However, it is at least implicitly accepted that teachers who are competent and equipped in terms of sustainable development will make significant contributions to the progress of students in this regard [72–74]. The most important here seems to be pedagogical knowledge [69]. This can be summarized as the ability to manage students' disruptive behavior in the classroom, personalize learning, and organize teaching processes [75].

Only a concerted and collective effort can transform a country or the world for the better (sustainability). In addition, societies and people are in constant interaction [64]. In this sense, the communication of societies on a global scale will also provide positive or negative support to this process. From an indirect and seemingly unrelated perspective, it can be said that it is a basic assumption that all societies should look at sustainability in terms of the environment as a higher quality of human life, stripping away their identities and prejudices. In this process, conducting education and all kinds of communication tools (social or traditional media) based on this perspective will make a positive contribution to the global and collective effort. Hydroelectric power plants are important energy sources in terms of renewable energy and therefore sustainable development [76]. Many developing countries have only been able to use half of their potential in small-scale hydroelectric power plants [77]. However, a comprehensive environmental impact assessment before the establishment of these plants is critical in terms of the sustainability of the plant and eliminating its negative effects on biodiversity [78]. However, due to reasons such as economic and political instabilities and lack of a culture of objection, a comprehensive environmental impact assessment cannot be conducted [79]. Environmental education offers a way to ensure positive environmental change [80]. In order to achieve the goal of positive environmental change, more holistic, collective and life-affecting changes can be made by providing effective environmental education to every unit of society [81]. In particular, individual behaviors have been primarily focused on solutions to global environmental problems such as climate change. However, many studies consider it critical to be able to put effective pressure on governments (in taking steps toward sustainable development) to solve global environmental problems such as preventing climate change. Here, it is implied that activities that prevent the breakdown of collective consciousness in societies will be useful [82]. In this sense, supporting a culture of objection and increasing freedom of thought seem to be useful.

Acknowledgements

I would like to thank all the members of my family who gave meaning to my life.

Conflict of interest

The author declares no conflict of interest.

Author details

Süleyman Akçay
Faculty of Education, Süleyman Demirel University, Department of Mathematics and Science Education, Isparta, Turkey

*Address all correspondence to: suleymanakcay@sdu.edu.tr

References

[1] Farhani S, Ben Rejeb J. Link between economic growth and energy consumption in over 90 countries. Interdisciplinary Journal of Contemporary Research in Business (IJCRB). 2012;**3**(11):282-297

[2] Yalçın M. Fossil fuel consumption, renewable energy and economic growth [unpublished master's thesis]. Karaman, Turkey: Karamanoğlu Mehmetbey University; 2022

[3] Ali M, Seraj M. Nexus between energy consumption and carbon dioxide emission: Evidence from 10 highest fossil fuel and 10 highest renewable energy-using economies. Environmental Science and Pollution Research. 2022;**29**(58):87901-87922. DOI: 10.1007/s11356-022-21900-9

[4] Khan MK, Khan MI, Rehan M. The relationship between energy consumption, economic growth and carbon dioxide emissions in Pakistan. Financial Innovation. 2020;**6**:1-13. DOI: 10.1186/s40854-019-0162-0

[5] Rani T, Wang F, Rauf F, Ain QU, Ali H. Linking personal remittance and fossil fuels energy consumption to environmental degradation: Evidence from all SAARC countries. Environment, Development and Sustainability. 2023;**25**(8):8447-8468. DOI: 10.1007/s10668-022-02407-2

[6] Statistical Review of World Energy. Statistical Review of World Energy. 2024. Available from: https://www.energyinst.org/statistical-review [Accessed: December 08, 2023]

[7] Satır Reyhan A, Reyhan H. Küresel ısınmanın nedenleri, sonuçları, çözümleri üzerine yeni değerlendirmeler. Memleket Siyaset Yönetim. 2016;**11**(26):1-24

[8] Atmospheric Carbon Dioxide Amounts and Annual Emissions 1950-2021. Interactions with the Atmosphere. 2024. Available from: https://scied.ucar.edu/learning-zone/earth-system/anthropocene [Accessed: December 08, 2024]

[9] Sinn HW. Yeşil Paradoks: Küresel Isınmaya Arz Yanlı Yaklaşım. (Çev. M. E. Dinçer). İstanbul: Koç Üniversitesi Yayınları; 2016

[10] Hudson P. The Industrial Revolution. New York: Oxford University Press; 2014

[11] Ashton TS. The Industrial Revolution 1760-1830. London: Oxford University Press; 1997

[12] Berg M, Hudson P. Rehabilitating the industrial revolution 1. The Economic History Review. 1992;**45**(1):24-50

[13] Albritton Jonsson F. The industrial revolution in the Anthropocene. The Journal of Modern History. 2012;**84**(3):679-696. DOI: 10.1086/666049

[14] Anthropocene. What is the Anthropocene? 2024. Available from: https://scied.ucar.edu/learning-zone/earth-system/anthropocene [Accessed: December 08, 2024]

[15] Worster D, editor. The Ends of the Earth: Perspectives on Modern Environmental History. New York: Cambridge University Press; 1988

[16] Jeswani H, Krüger C, Russ M, Horlacher M, Antony F, Hann S, et al. Life cycle environmental impacts of

chemical recycling via pyrolysis of mixed plastic waste in comparison with mechanical recycling and energy recovery. Science of the Total Environment. 2021;**769**:144483. DOI: 10.1016/j.scitotenv.2020.144483

[17] Avan C. Determination of students' attitudes about the recycling of plastic and plastic wastes and their effects on the environment [unpublished master's theses]. Kastamonu, Turkey: Kastamonu University; 2011

[18] Keleş O. Implementation and evaluation of ecological footprint as an environmental education tool for sustainable living [unpublished doctoral theses]. Ankara, Turkey: Gazi University; 2007

[19] Can NS. Investigation of out-of-school learning environment activities in terms of different variables for primary school students about recycling and its effects on the environment [unpublished master's theses]. Erzincan, Turkey: Erzincan Binali Yıldırım University; 2019

[20] Yiğit K. The effect of recycling education for sustainable life on environmental awareness of 8th grade students [unpublished master's theses]. Istanbul, Turkey: Marmara University; 2019

[21] Hopper JR, Nielsen JM. Recycling as altruistic behavior: Normative and behavioral strategies to expand participation in a community recycling program. Environment and Behavior. 1991;**23**(2):195-220. DOI: 10.1177/0013916591232004

[22] Kaya A. Determining the views of science teachers about recycling [unpublished master's theses]. Giresun, Turkey: Giresun University; 2017

[23] Harris PG, editor. Routledge Handbook of Global Environmental Politics (No. s 31). New York: Routledge; 2014

[24] Chasek PS, Downie DL, Brown JW. The emergence of global environmental politics. In: Chasek PS, Downie DI, Brown JW, editors. Global environmental politics. London: Routledge; 2006. pp. 1-16

[25] Clark CR. Collective action competence: An asset to campus sustainability. International Journal of Sustainability in Higher Education. 2016;**17**(4):559-578. DOI: 10.1108/IJSHE-04-2015-0073

[26] Kenis A, Mathijs E. Beyond individual behaviour change: The role of power, knowledge and strategy in tackling climate change. Environmental Education Research. 2012;**18**(1):45-65. DOI: 10.1080/13504622.2011.576315

[27] Davis L, Fähnrich B, Nepote AC, Riedlinger M, Trench B. Environmental communication and science communication—Conversations, connections and collaborations. Environmental Communication. 2018;**12**(4):431-437

[28] Monroe MC, Andrews E, Biedenweg K. A framework for environmental education strategies. Applied Environmental Education and Communication. 2008;**6**(3-4):205-216. DOI: 10.1080/15330150801944416

[29] Wibeck V. Enhancing learning, communication and public engagement about climate change–some lessons from recent literature. Environmental Education Research. 2014;**20**(3):387-411. DOI: 10.1080/13504622.2013.812720

[30] World Commission on Environment and Development (WCED). Our Common Future—The Bruntland Report. Oxford: Oxford University Press; 1987

[31] Scoullos M. Towards an environmental education for sustainable development. In: United Nations Educational, Scientific, and Cultural Organization (UNESCO), Interregional Workshop on Re-orienting Environmental Education for Sustainable Development; 26-30 June, Annex 6. Athens: UNESCO; 1995. pp. 1-10

[32] Leal Filho W, Manolas E, Pace P. The future we want: Key issues on sustainable development in higher education after Rio and the UN decade of education for sustainable development. International Journal of Sustainability in Higher Education. 2015;**16**(1):112-129. DOI: 10.1108/IJSHE-03-2014-0036

[33] Acosta Castellanos PM, Queiruga-Dios A. From environmental education to education for sustainable development in higher education: A systematic review. International Journal of Sustainability in Higher Education. 2022;**23**(3):622-644. DOI: 10.1108/IJSHE-04-2021-0167

[34] Tilbury D. Environmental education for sustainability: Defining the new focus of environmental education in the 1990s. Environmental Education Research. 1995;**1**(2):195-212. DOI: 10.1080/1350462950010206

[35] McBride BB, Brewer CA, Berkowitz AR, Borrie WT. Environmental literacy, ecological literacy, ecoliteracy: What do we mean and how did we get here? Ecosphere. 2013;**4**(5):1-20

[36] Roth CE. Curriculum Overview for Developing Environmentally Literate Citizens. 1968. Available from: https://files.eric.ed.gov/fulltext/ED032982.pdf

[37] Roth CE. A Questioning Framework for Shaping Environmental Literacy (US, Earthlore Associates ve the Center for Environmental Education of Antioch New England Institute). 2002. Available from: https://files.eric.ed.gov/fulltext/ED348235.pdf

[38] Carson R. Silent Spring. New York: Houghton-Mifflin; 1962

[39] Rothman HK. The Greening of a Nation? Environmentalism in the United States since 1945. Orlando: Harcourt Brace; 1998

[40] Stapleton SR. Environmental identity development through social interactions, action, and recognition. The Journal of Environmental Education. 2015;**46**(2):94-113. DOI: 10.1080/00958964.2014.1000813

[41] Clayton SD, editor. The Oxford Handbook of Environmental and Conservation Psychology. New York: Oxford University Press; 2012

[42] Kopnina H. Education for the future? Critical evaluation of education for sustainable development goals. The Journal of Environmental Education. 2020;**51**(4):280-291. DOI: 10.1080/00958964.2019.1710444

[43] Vernadsky VI. The Biosphere (D. B. Langmuir, Trans.). New York: Springer; 1998

[44] Office of Biological and Environmental Research of the U.S. Department of Energy Office of Science. Carbon Cycle Diagram from the DOE with Numbers. 2024. Available from: https://scied.ucar.edu/image/carbon-cycle-diagram-doe-numbers [Accessed: December 08, 2024]

[45] Capra F. Sustainable living, ecological literacy, and the breath of life. Canadian Journal of Environmental Education (CJEE). 2007;**12**(1):9-19

[46] Mohan L, Chen J, Anderson CW. Developing a multi-year learning

progression for carbon cycling in socio-ecological systems. Journal of Research in Science Teaching: The Official Journal of the National Association for Research in Science Teaching. 2009;**46**(6):675-698

[47] Hartley LM, Wilke BJ, Schramm JW, D'Avanzo C, Anderson CW. College students' understanding of the carbon cycle: Contrasting principle-based and informal reasoning. Bioscience. 2011;**61**(1):65-75. DOI: 10.1525/bio.2011.61.1.12

[48] Akçay S. Prospective elementary science teachers' understanding of photosynthesis and cellular respiration in the context of multiple biological levels as nested systems. Journal of Biological Education. 2016;**51**(1):52-65. DOI: 10.1080/00219266.2016.1170067

[49] Skribe Dimec D, Strgar J. Scientific conceptions of photosynthesis among primary school pupils and student teachers of biology. CEPS Journal. 2017;**7**(1):49-68. DOI: 10.25656/01:12958

[50] Bielak AT, Campbell A, Pope S, Schaefer K, Shaxson L. From science communication to knowledge brokering: The shift from 'science push' to 'policy pull'. In: Cheng D, Claessens M, Gascoigne T, Metcalfe J, Schiele B, Shi S, editors. Communicating Science in Social Contexts: New Models, New Practices. New York: Springer; 2008. pp. 201-226

[51] Likens GE. The role of science in decision making: Does evidence-based science drive environmental policy? Frontiers in Ecology and the Environment. 2010;**8**(6):e1-e9. DOI: 10.1890/090132

[52] Elegbede I, Matti-Sanni R, Moriam O, Emily Osa I. Sustainability education and environmental awareness. In: Idowu S, Schmidpeter R, Capaldi N, Zu L, Del Baldo M, Abreu R, editors. Encyclopedia of Sustainable Management. Cham: Springer; 2023. DOI: 10.1007/978-3-030-02006-4_128-1

[53] Akdemir MA, Bal CG. Sustainability communication in higher education institutions: Scale development and validation study. OPUS Journal of Society Research. 2024;**21**(5):257-273. DOI: 10.26466/opusjsr.1511496

[54] Pitton V, McKenzie M. What moves us also moves policy: The role of affect in mobilizing education policy on sustainability. Journal of Education Policy. 2020;**37**(4):527-547. DOI: 10.1080/02680939.2020.1852605

[55] Rickinson M, McKenzie M. The research-policy relationship in environmental and sustainability education. Environmental Education Research. 2021;**27**(4):465-479. DOI: 10.1080/13504622.2021.1895973

[56] Esteban Ibáñez M, Musitu Ferrer D, Amador Muñoz LV, Claros FM, Olmedo Ruiz FJ. University as change manager of attitudes towards environment (the importance of environmental education). Sustainability. 2020;**12**(11):4568. DOI: 10.3390/su12114568

[57] Pihkala P. Eco-Anxiety and environmental education. Sustainability. 2020;**12**(23):10149. DOI: 10.3390/su122310149

[58] Aleixo AM, Leal S, Azeiteiro UM. Higher education students' perceptions of sustainable development in Portugal. Journal of Cleaner Production. 2021;**327**:129429. DOI: 10.1016/j.jclepro.2021.129429

[59] Leal S, Azeiteiro UM, Aleixo AM. Sustainable development in Portuguese higher education institutions from the faculty perspective. Journal of

Cleaner Production. 2024;**434**:139863. DOI: 10.1016/j.jclepro.2023.139863

[60] Owojori OM, Mulaudzi R, Edokpayi JN. Student's knowledge, attitude, and perception (KAP) to solid waste management: A survey towards a more circular economy from a rural-based tertiary institution in South Africa. Sustainability. 2022;**14**(3):1310. DOI: 10.3390/su14031310

[61] Jančius R, Gavenauskas A. The influence of values and social environment on parents' environmental attitudes: Lithuanian case study. Sustainability. 2022;**14**(20):13415. DOI: 10.3390/su142013415

[62] Orbanić ND, Kovač N. Environmental awareness, attitudes, and behaviour of preservice preschool and primary school teachers. Journal of Baltic Science Education. 2021;**20**(3):373. DOI: 10.33225/jbse/21.20.373

[63] Debrah JK, Vidal DG, Dinis MAP. Raising awareness on solid waste management through formal education for sustainability: A developing countries evidence review. Recycling. 2021;**6**(1):6. DOI: 10.3390/recycling6010006

[64] Raghu SJ, Rodrigues LLR. Behavioral aspects of solid waste management: A systematic review. Journal of the Air and Waste Management Association. 2020;**70**(12):1268-1302. DOI: 10.1080/10962247.2020.1823524

[65] Ziadat AH. Major factors contributing to environmental awareness among people in a third world country/Jordan. Environment, Development and Sustainability. 2010;**12**:135-145. DOI: 10.1007/s10668-009-9185-4

[66] Abolore AA. Comparative study of environmental sustainability in building construction in Nigeria and Malaysia. Journal of Emerging Trends in Economics and Management Sciences. 2012;**3**(6):951-961

[67] N'dri LM, Islam M, Kakinaka M. ICT and environmental sustainability: Any differences in developing countries? Journal of Cleaner Production. 2021;**297**:126642. DOI: 10.1016/j.jclepro.2021.126642

[68] Sleurs W. Competencies for ESD (Education for Sustainable Development) Teachers. A Framework to Integrate ESD in the Curriculum of Teacher Training Institutes: Comenius 2.1 Project 118277-CP-1-2004-BE-Comenius-C2, 1. 2008. Available from: http://www.unece.org/fileadmin/DAM/env/esd/inf.meeting.docs/EGonInd/8mtg/CSCT%20Handbook_Extract.pdf

[69] Bertschy F, Künzli C, Lehmann M. Teachers' competencies for the implementation of educational offers in the field of education for sustainable development. Sustainability. 2013;**5**(12):5067-5080. DOI: 10.3390/su5125067

[70] Kadji-Beltran C, Zachariou A, Stevenson RB. Leading sustainable schools: Exploring the role of primary school principals. Environmental Education Research. 2013;**19**(3):303-323. DOI: 10.1080/13504622.2012.692770

[71] Vare P. A rounder sense of purpose: Developing and assessing competences for educators of sustainable development. Form@re. 2018;**18**(2):164-173. DOI: 10.13128/formare-23712

[72] Brandt J-O, Bürgener L, Barth M, Redman A. Becoming a competent teacher in education for sustainable development: Learning outcomes and processes in teacher education.

International Journal of Sustainability in Higher Education. 2019;**20**(4):630-653. DOI: 10.1108/IJSHE-10-2018-0183

[73] Imara K, Altinay F. Integrating education for sustainable development competencies in teacher education. Sustainability. 2021;**13**(22):12555. DOI: 10.3390/su132212555

[74] Mohammadi M, Moradi K. Exploring change in EFL teachers' perceptions of professional development. Journal of Teacher Education for Sustainability. 2017;**19**(1):22-42. DOI: 10.1515/jtes-2017-0002

[75] Manasia L, Ianos MG, Chicioreanu TD. Pre-service teacher preparedness for fostering education for sustainable development: An empirical analysis of central dimensions of teaching readiness. Sustainability. 2020;**12**(1):166. DOI: 10.3390/su12010166

[76] Jadoon TR, Ali MK, Hussain S, Wasim A, Jahanzaib M. Sustaining power production in hydropower stations of developing countries. Sustainable Energy Technologies and Assessments. 2020;**37**:100637. DOI: 10.1016/j.seta.2020.100637

[77] Hennig T, Harlan T. Shades of green energy: Geographies of small hydropower in Yunnan, China and the challenges of over-development. Global Environmental Change. 2018;**49**:116-128. DOI: 10.1016/j.gloenvcha.2017.10.010

[78] Couto TB, Olden JD. Global proliferation of small hydropower plants–science and policy. Frontiers in Ecology and the Environment. 2018;**16**(2):91-100. DOI: 10.1002/fee.1746

[79] Kuriqi A, Pinheiro AN, Sordo-Ward A, Bejarano MD, Garrote L. Ecological impacts of run-of-river hydropower plants—Current status and future prospects on the brink of energy transition. Renewable and Sustainable Energy Reviews. 2021;**142**:110833. DOI: 10.1016/j.rser.2021.110833

[80] Ardoin NM, Bowers AW, Gaillard E. Environmental education outcomes for conservation: A systematic review. Biological Conservation. 2020;**241**:108224. DOI: 10.1016/j.biocon.2019.108224

[81] Ardoin NM, Heimlich JE. Environmental learning in everyday life: Foundations of meaning and a context for change. Environmental Education Research. 2021;**27**(12):1681-1699. DOI: 10.1080/13504622.2021.1992354

[82] Gibson L, Busch KC, Stevenson K, Chesnut L, Cutts B, Seekamp E. Conceptualizing community-level environmental literacy using the Delphi method. Environmental Education Research. (In press):1-30. DOI: 10.1080/13504622.2024.2397595

Chapter 2

An Assessment of the Risks Relating to Carbon Finance Investments Policies for Recovery, Adaptation, and Mitigation of Climate Change Impacts

Nabulya Jesca, Peter Oseghale Ohue and Jerome Ntege

Abstract

Financial growth mitigates climate change impacts and risks calling for policies to mitigate, adapt, and recover from climate change impacts. Despite interventions like carbon finance investments (CFIs), the net zero emission (NZE) goal has not been met. This chapter aimed to assess the risks of CFI policies to guide decision-making. CFIs involve a reduction of greenhouse gas, generating returns on carbon markets, and supporting sustainable development. The study employed a secondary research design using data from the World Bank, International Energy Agency (IEA), and Scopus from 2009 to June 2024. Data was analyzed using thematic analysis and Ordinary Least Squares (OLS). The study found a conflict between the social and business dilemma and the high concentration of CFIs in advanced economies. It suggests balanced investments toward sectors and regions. In conclusion, the low funding and capacity to manage resources limit the scalability and size of CFIs calling for collaboration, coordination, and resource commitment.

Keywords: risks, carbon finance investments, climate change, recovery, adaptation, mitigation, sustainable development, green bonds, carbon credit, net zero emission, climate finance

1. Introduction

Climate change is a global reality rising out of economic growth with concentrations of greenhouse gases [1], a threat that has confronted many countries and reversed years of progress [2]. Climate change directly threatens the well-being of humankind including loss of life with a future full of uncertainties [3]. The processes of mitigation, adaptation, and recovery involve funding strategies and mechanisms to address the challenges and risks associated with climate change [1, 4, 5] which requires stakeholders' participation in carbon markets. Carbon Finance Investments (CFIs) policies play a vital role in recovering, adapting, and mitigating climate change impacts [5, 6]. This chapter presents the definition of CFIs and the justification of CFIs policies in addressing impacts of climate change. The chapter focuses on economic, social, political, and economic threats and opportunities to CFIs. This is done in line with the associated goals of net zero emission (NZE), carbon credit sales to generate returns to investors [5], and protecting the natural ecosystem. It assesses the strategies that should be globally pursued through CFIs to achieve sustainable development goals [7, 8].

The United Nations Framework Convention on Climate Change (UNFCCC) officially defines climate finance as finance flowing directly to assets and activities aiming at reducing emissions and enhancing sinks of greenhouse gases. This aims at reducing the vulnerability of, maintaining, and increasing the resilience of human and ecology systems to negative climate change impacts [9]. Carbon finance investments involve the commitment of resources by both the governments and the private sector in projects. It also involves initiatives aimed at supporting climate change mitigation and its impacts [1, 4, 10]. CFIs are guided by the goals of reducing greenhouse gas emissions [1, 4, 6, 10] mainly carbon dioxide. CFIs generate returns through carbon credit sales for the project developers on carbon markets such as green bonds [4, 5, 10], and support sustainable development through energy efficient improvement and promoting the use of renewable energy [1, 4–6, 10, 11]. CFIs enhance environmental protection through forestry and land use projects, promote industrial processes toward NZE [12, 13], proper waste management, and methane capture [4].

Carbon finance investment interventions such as green bonds, carbon offsetting programs, and carbon credit have been found to address the impacts of climate change and promote the achievement of sustainable development goals [4–6]. CFIs provide opportunities for supporting basic development needs such as energy access with the highest volume of investments. This is done toward mitigation as compared to adaptation as the saying goes "prevention is better than cure" to address the impacts of climate change [10, 12]. There is increased technology transfer and diffusion [4], growth in funding by investors, and availability of global resources funded by governments and the private sector (**Figure 1**) [1, 4, 10]. In addition, there is an increased global awareness which eases progress toward an ecologically sustainable future, financing ecofriendly projects and enabling a transition away from fossil fuels [4, 5]. This is provided a financial facility to invest in clean and renewable energy to maintain a natural resource-based sustainable future [5, 6, 12], creating carbon neutrality [6], and promoting social equity [5].

CFIs have brought together different stakeholders which has increased networking among governments such as the G4, G20, and the private sector [1, 12]. This along with the multilateral agencies sets clear targets relating to climate change such as the Zero Carbon Emission by 2050 and the achievement of sustainable development goals by 2030 [5, 12]. The growth of carbon markets led to increased commitment and collaborations [4], providing a platform for participation in the capital markets to address the causes of climate change and its impact. All of these give direction and checkpoints concerning adherence to climate policies and interventions in place despite a number of threats.

The widening income gap has been identified as one of the threats to scalability of CFIs [1, 10]. With the large wealth disparity that necessitates investment from the developed world to the LDCs, there is a glaring inability to adapt to low-carbon technology to facilitate the transition to a climate-resilient economy [4]. Successful

DOI: http://dx.doi.org/10.5772/intechopen.1007154

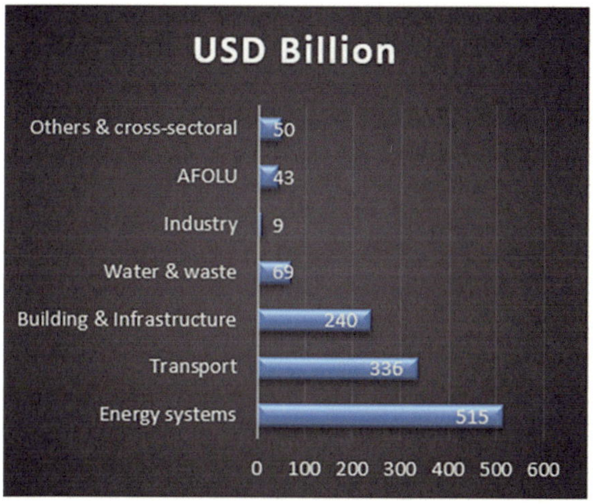

Figure 1.
Global investments in energy transition technologies for the year 2023. Source: Climate Policy Initiative [8].

implementation of CFI is limited by the poor coordination among stakeholders both at the local and international levels [2, 4]. The other factors include the geopolitical tension among sovereign independent states with greater national roles and responsibilities [4]. Political wars and the regulatory risks to project validation and registration processes are the remaining factors [4]. The debt burdens especially in LDC put the CFIs at a great risk of default given the mere fact that most of the investments are being financed by debt as compared to equity capital [4]. This limits the size of investments due to the high risks assessed by the would-be investors. Further, the growing income inequality gap and digital divide at the different economic levels limit individuals' participation in CFIs. Other threats include international diplomacy [1, 9], the high population growth rate especially in sub-Saharan Africa [2] which increases pressure on the available resources for developing economies, the high poverty, and poor culture all limit channel resources toward CFI to promote sustainable development [4].

The increase in dependence and a culture of laziness in the developing world limits participation in climate change mitigation, adaptation, and recovery activities including undertaking investments in carbon finance. In this case, most Emerging Markets and Developing Economies (EMDEs) have to wait for financial and physical handouts instead of being innovative to find alternatives. These alternatives could be used to address their challenges which are related to climate change such as social distress, massive famine outbreaks, school dropouts, increased migration, and health risks. Given the poor capacity to manage CFI projects, LDCs have not been able to tap from the global resources available so that they can meet the costs of climate change. Addressing climate change effectively requires substantial financial investments [10].

Figure 1 shows the global investments in energy transition technologies for the year 2023. According to the Climate Policy Initiative, the global climate finance flow for 2021 and 2022 averaged 1.27 trillion USD annually. **Figure 1** shows the breakdown of this finance channeled into various activities, including adaptation (63 billion USD), dual benefits (51 billion USD), and mitigation (1.15 trillion USD). These funds were allocated across several sectors: energy systems (515 billion USD), transport (336 billion USD), building and infrastructure (240 billion USD), water and waste

(69 billion USD), industry (9 billion USD), agriculture, forestry, other land use, and fisheries (AFOLU) (43 billion USD), and other cross-sectoral activities (50 billion USD). Between 2011 and 2023, there has been an upward trend from 394 billion USD to over 1 trillion USD [11] in line with mitigation, adaptation, and recovery of climate change impacts. The highest portion of climate finance in 2021–2022 was allocated toward mitigation efforts (**Table 1** and **Figure 2**). Energy systems received the largest share of finance, followed by transport and building and infrastructure. Smaller amounts were allocated to water and waste, industry, AFOLU, and other cross-sectoral activities which increase the risks to mitigate, adapt, and recover from climate change impacts.

In as much as the available information mainly focuses on the risks associated with climate change such as global warming, migrations, social stress, famine, and the natural environment, less attention is given to the risks associated with carbon finance investments. The high level of inconsistency in CFIs and fossil fuels investments has resulted in planning complications and the management of the entire system to address climate change impacts. The timing and scalability of CFIs are limited particularly in EMDEs, attributed to the rapid pace of population increase and the rising need for cooking fuel, geopolitical tensions, and the lack of coordination. The increased use of fossil fuels and the dominance of the private sector for climate finance driven by the profit motive and understanding of climate change efforts particularly among the impoverished people [14–16], all lead to the increase in carbon emission.

Despite the intensity of the mitigation effort regarding carbon emissions, there is still a huge disparity between the actions, the volume of emissions, and the ambition of limiting global warming to 1.5°C. The increase in renewable energy investments

Type of finance	USD billion
Adaptation	63
Dual benefits	51
Mitigation	1150
Source: Climate Policy Initiative [8].	

Table 1.
Global investments in adaptation, dual benefits, and mitigation of climate change impacts for the year 2023.

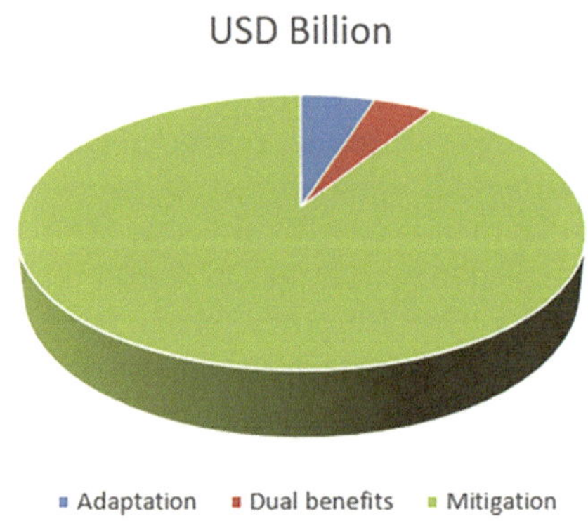

Figure 2.
Visualization of the global investments in adaptation, dual benefits, and mitigation of climate change impacts for the year 2023. Source: Climate Policy Initiative [8].

outpaced the rate of investments in fossil fuels as of 2023. Also, fossil fuel investments that account for the highest level of greenhouse gas emissions more than doubled the levels required to meet the much lower demand in the Net Zero emissions scenario [12, 17]. A related and much broader concept is the 'consistency' or 'alignment' of financial flows with global financial goals [18].

There is a mismatch between finance for real activities on the ground and activity in the financial markets—prompting a wide range of activities, both public and private, in the financial sector [1]. Further, there was unbalanced funding through CFIs to the targeted sectors that would reduce carbon emissions and promote mitigation, adaptation, and recovery to the impacts of climate change most especially in the developing countries. These countries are the most vulnerable to the impacts of climate change with only US$16.5 billion representing less than 2.5% of the total investments in carbon finance investments. Funding that went to energy was up to US$336 billion per year on average and sustainable transport with US$169 billion in 2019–2020 (**Figure 2**) [18]. This calls for strategic interventions relating to the management of CFIs.

The increase in CFI engagement and involvement of the developing world in the planning phase of climate change interventions is one strategy that helps cut carbon emissions and address mitigation, recovery, and adaptation to climate change. Instead of designating a class of receivers and donors, the involvement of developing economies increases participation and commitment to mitigation and addressing the effects of climate change in a sustainable way. This can be achieved by increasing CFIs in the leading sectors for EMDEs and promoting the use of local resources to encourage local innovations in managing climate change impacts. Issues relating to cooking fuel which has been identified to be a major challenge in mitigating and addressing the impacts of climate change should be handled by empowering local communities based on the available innovations and resources. Innovations such as the use of bamboo for cooking should be targeted by CFIs since it is much more accessible and cheaper. They are also more ecofriendly compared to programs such as rural electrification where electric poles pass on the rooftops of households.

There is a need for economic freedom in poor societies if the global community is to mitigate and control the impacts of climate change through CFIs. CFIs should target individual projects by households and Small and Medium Scale Enterprises (SMEs) through empowerment. This is better than the huge capital investments like infrastructure development which do not directly involve the participation of households. This typically hinders achieving CFI goals and the realization of sustainable development goals.

2. Carbon finance investments (CFIs) risks on mitigation, recovery, and adaptation to climate change impacts

Risks associated with CFIs include; social, political, economic, and environmental opportunities and threats (**Figure 3**). There is a strong alignment of climate and energy security goals and a focus on industrial strategy as countries seek to strengthen their footholds in the emerging clean energy economy [17]. However, CFIs have targeted basically large-scale manufacturers based on the industrialization strategy to address the issues of climate change with less involvement by SMEs. Further, targeting activities by individual households such as affordable and accessible cooking fuel despite massive investments in clean energy have not yet yielded tangible results to reduce carbon emissions and protection of the ecosystem.

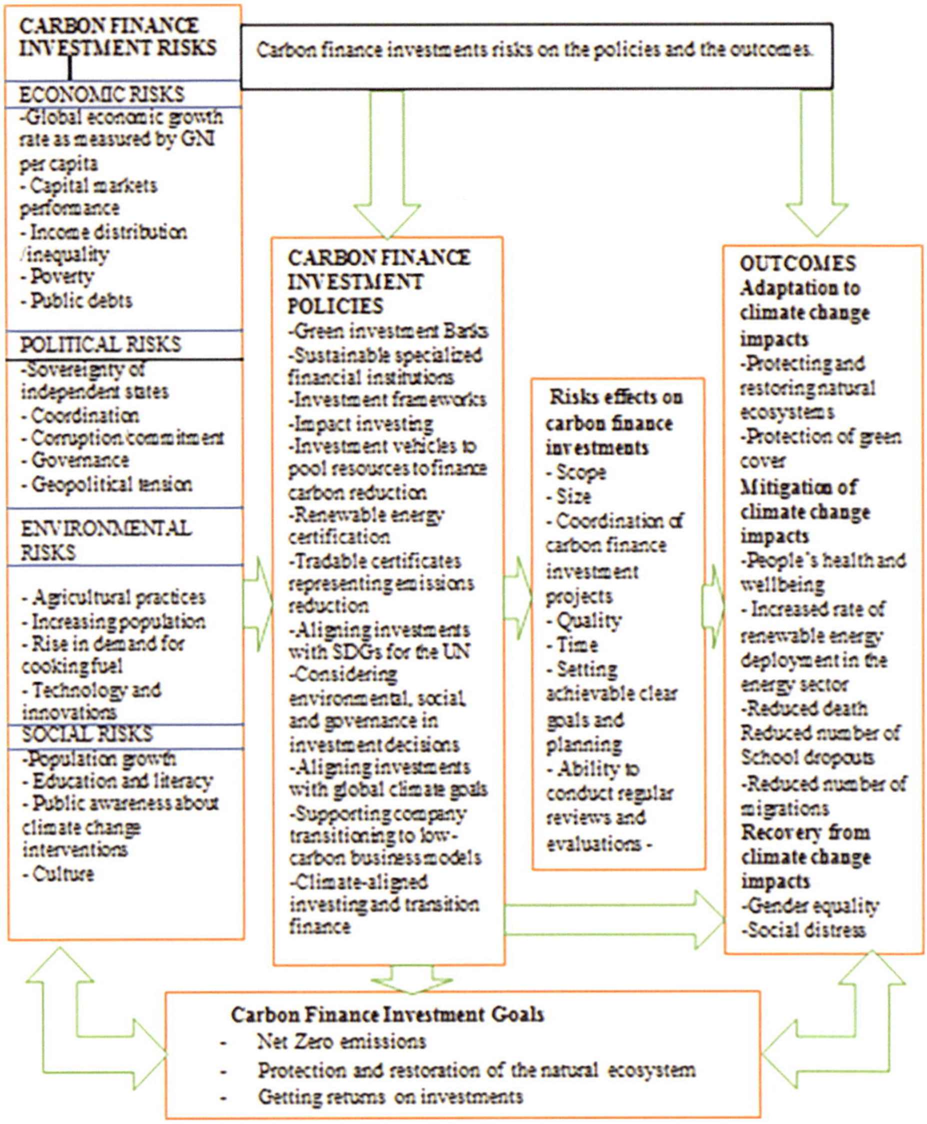

Figure 3.
A Causal relationship between carbon finance investments, identified risks, CFIs goals, how the risks affect carbon finance investments, and outcomes in terms of adaptation, Mitigation, and recovery from climate change impacts. Source: Designed by the researchers.

The daily need for cooking fuel has created challenges, particularly for the most vulnerable poor communities. These challenges have exacerbated the effects of climate change by destroying natural vegetation cover, which in turn destroys the ecosystem as a whole [10]. This is partly responsible for the action in 2022, when the global investments in energy transition technologies, including energy efficiency, reached \$1.3 trillion, a record high and a 19% increase from 2021 and a 70% increase from the pre-pandemic levels in 2019 [17]. However, investments in renewable energy which totaled \$0.5 trillion in 2022 were still less than one-third of average investments needed annually between 2023 and 2030 to achieve the 1.5°C Scenario. Statistics

DOI: http://dx.doi.org/10.5772/intechopen.1007154

show an increase in investments in fuel supply with an average above $500 billion in upstream oil and gas, $200 billion in midstream oil and gas, $50 billion average in coal, and less than $10 billion average in low-emission fuels from 2019 to 2023 [17]. Though importing technology and inventions from large firms that are environmentally friendly has been pursued by advanced economies, such energy initiatives cannot be enrolled, especially among the poor.

Statistically, the largest percentage of investments is directed toward mitigation measures compared to adaptation which justifies the heavy investments in energy and infrastructure. There is an imbalance in funding based on needs as LDCs; the most vulnerable to climate change impacts required to adopt yet the advanced nations are trying out possibilities on how they can control climate change impacts. CFIs have been challenged by a lack of capacity in project preparations and the limited replicability and scalability of matchmaking interventions [11].

2.1 Economic risks of carbon finance investments policies and climate change outcomes

Economic risks involve the threats and opportunities to CFIs based mainly on the availability and efficiency in the utilization of the resources in order to mitigate and address the impacts of climate change. This includes the global economic trends (recessions and growth rate), financial performance in carbon markets, income distribution, poverty, public debts, and the level of international trade. These further involve government policy support through instruments such as the US Inflation Reduction Act, and new initiatives in Europe, Japan, and China [17]. Economic risks account for the availability of resources (**Figure 4**), scalability, plus the timely response to climate change impacts.

Figure 4 shows an escalation in the investments in global climate finance recorded on a biennial basis. This serves as additional motivation for this exploration and

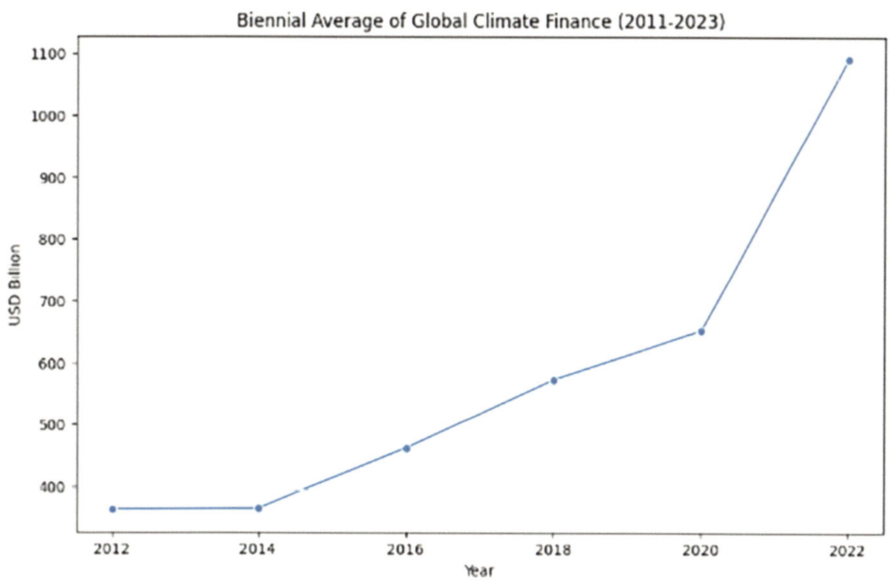

Figure 4.
Plot of the growth biennial average of global climate finance. Source: Climate Policy Initiative [8].

projections. However, CFIs are characterized by unforeseen challenges [4] for those investors who want to build their portfolio based on carbon finance projects such as the longer carbon finance revenue streams needed which impacts demand for carbon credits [4]. The long and unpredictable project cycles make it difficult to estimate project viability yet the payments are based on performance [4]. Over time, this reduces the pool of possible investors rather than making the projects more appealing to broaden the pool of investments in carbon finance. Other economic downturns exist as well, such as global recessions. Nevertheless, there are obstacles to the financial markets' effective operation. The high cost of capital, which further discourages both government and private investors in green energy alternatives, maybe the reason why innovators are less able to fulfill climate targets of lower carbon emissions. These risks, including inflation, also lead innovators to seek a higher rate of return from their investments despite the growing demand for resources to mitigate, adapt, and recover from climate change impacts.

CFIs face risks of funding for climate change impact and mitigation based on the available financial resources. CFIs have been funded basically by the private sector, governments, UN agencies such as IFC, bilateral, and multilateral channels (**Table 2** and **Figure 5**) [6]. The funding risks have been attributed to basically the ability to forecast the climate change-related needs to facilitate the planning process and timely resource mobilization in addressing the challenges of climate change (**Figure 6**). The assessors cannot predict with certainty the quantity of finance needed to deliver on climate change goals [9]. A cumulative amount of $600 billion per year was needed until 2030 to meet carbon finance investment needs by the developing countries, while another report estimated a total of $US1 trillion to tackle climate change and drive development in EMDEs excluding China by 2025 and around $US2.5 trillion per year from 2030 [9]. Despite the developed countries' commitment to funding developing countries with $100 billion annually to address climate change issues, there has been a noted lack of financial resources to adapt to climate change, particularly in the LDCs [9], who are more vulnerable to climate change impacts. The estimates for 2040, 2050, 2070, and even 2100 as shown in **Figure 6** indicate that varying sums of money will be required to meet climate targets.

Source	USD billion
Government	100
Corporations	192
Households and individuals	184
Commericial FIs	235
National DFIs	238
Bilateral DFIs	33
Multilateral DFIs	93
Multilateral climate funds	3
SOEs	110
State-owned FIs	61
Other	15

Table 2.
Investments distribution across sectors.

DOI: http://dx.doi.org/10.5772/intechopen.1007154

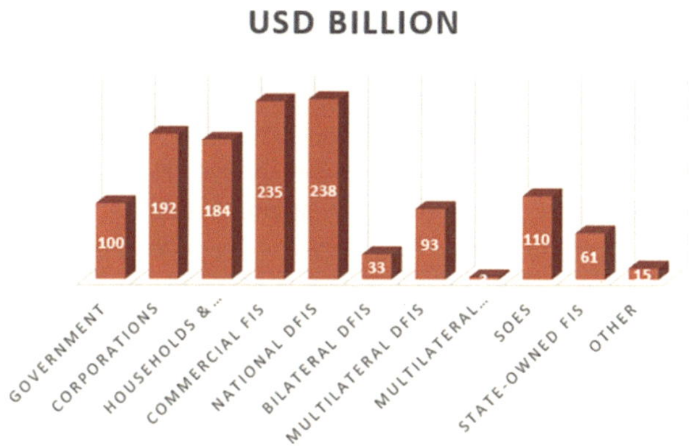

Figure 5.
Visualization of the investment distribution across sectors. Source: Climate Policy Initiative [8].

The comprehensive statistics and visualizations provided in this report show the present situation and future outlook for global climate finance. The ramifications of these results highlight the necessity of quicker and more equal financial flows to all sectors and regions in order to fund mitigation, adaptation, and recovery of climate change impacts. An Ordinary Least Squares (OLS) analysis and model diagnostics were carried out. The OLS regression analysis involves fitting a linear regression model to the data with year as the independent variable and finance in USD as the dependent variable. Beyond data collection, the model specification is given as:

$$\text{USD}(\text{Billion}) = \beta 0 + \beta 1.\text{Year} + \epsilon \tag{1}$$

Where β0, β1, and ε are the intercept, slope, and error terms respectively. The coefficients β0 and β1 are estimated using the least squares to minimize the sum of the squared residuals. With regard to diagnostics, the statistical metrics used are R-squared and Adjusted R-squared for measuring how well the model explains the variability of the dependent variable. F-statistics and its p-value are used for testing the overall significance of the model. Log-likelihood, Akaike information criterion (AIC), and Bayesian information criterion (BIC) metrics are for the goodness of fit and model comparison. To test the significance of individual predictors, t-statistics is used, Durbin-Watson statistics is used for testing autocorrelation of the residuals,

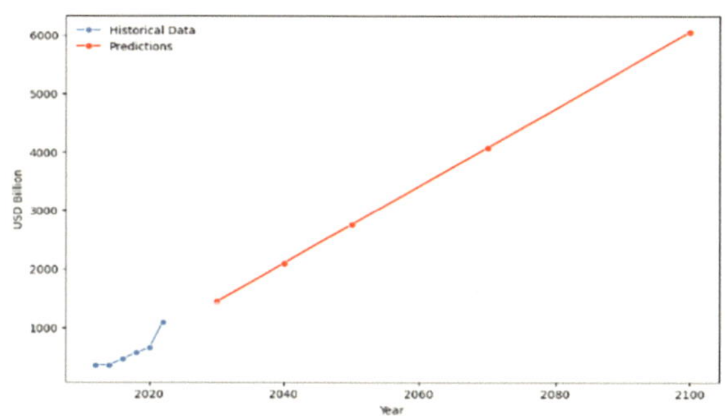

Figure 6.
Average climate finance (historical and projected). Source: Climate Policy Initiative.

and Jarque-Bera is used for testing for normality of the residuals. The summary table below shows the result of the model diagnostics (**Table 3**).

The model diagnostics above provide insights into the performance and validity of the statistical model used to analyze carbon financing. The standard errors assume that the covariance matrix of the errors is correctly specified. The condition number is large, 1.19e + 06 suggesting the presence of strong multicollinearity among the predictors. The R-squared indicates that 81.4% of the variability in carbon financing, confirms the model's effectiveness in capturing key trends and relationships. The model's statistical significance is supported by the p-value (0.0139), which is below the threshold of 0.05, suggesting the results are unlikely due to random chance. The Jarque-Bera test result of 0.725 shows that the residuals are normally distributed, affirming the model's assumptions and its overall validity in analyzing the factors influencing carbon financing.

Fit and complexity are balanced when evaluating and comparing statistical models using the AIC and the BIC measures. Whereas AIC enforces a penalty on the number of parameters to take care of overfitting favoring models with smaller values, simpler models use BIC penalization of model complexity as sample size increases.

OLS regression results						
Dep. Variable:			USD (Billion)	R-squared:		0.814
Model:			OLS Adj.	R-squared:		0.767
Method:			Least Squares	F-statistic:		17.47
Date:			Tue, 16 Jul 2024	Prob (F-statistic):		0.0139
Time:			11:05:56	Log-Likelihood:		−36.592
No. Observations:			6	AIC:		77.18
Df Residuals:			4	BIC:		76.77
Df Model:			1			
Covariance Type:			nonrobust			
	Coef	Std err	t	P > |t|	[0.025	0.975]
Const	−1.324e+05	3.18e+04	−4.161	0.014	−2.21e+05	−4.41e+04
Year	65.9286	15.773	4.180	0.014	22.136	109.721
Omnibus:			nan	Durbin-Watson:		1.663
Prob (Omnibus):			nan	Jarque-Bera (JB):		0.644
Skew:			0.533	Prob (JB):		0.725
Kurtosis:			1.799	Cond. No.		1.19e+06

Table 3.
OLS regression results.

The model fits the data well and takes into consideration its complexity with the significant value of AIC (77.18) and BIC (76.77).

To satisfy the growing demand for funds to address the impacts of climate change, the globe is not investing enough. One way out of this deadlock is a push to accelerate the deployment of clean energy technology and infrastructure, but this must happen soon enough to avoid more instability and volatility in the world energy markets [17]. The unequal development of the developed and developing worlds has further impacted the CFIs, with supply chains and skills being identified as potential bottlenecks that could limit their ability to flourish [17]. To expand the size and scale of CFI into clean energy transitions which will directly help the efforts to cut greenhouse gas emissions (**Figure 7**), it is required to mobilize significantly bigger support from development financing institutions.

2.2 The political risks of carbon finance investment policies on the outcomes of climate change

Political risks involve the commitment by independent states to get directly concerned and participate in climate-related activities. Governments should set clear terms, policies, and regulations, mobilizing and directing resource usage by the stakeholders. There is a need to enforce climate-related decisions with the right governance structures in place [6]. The rise in political uncertainty increases the risks of financial securities and the ability of corporations to have clear business strategies which are in

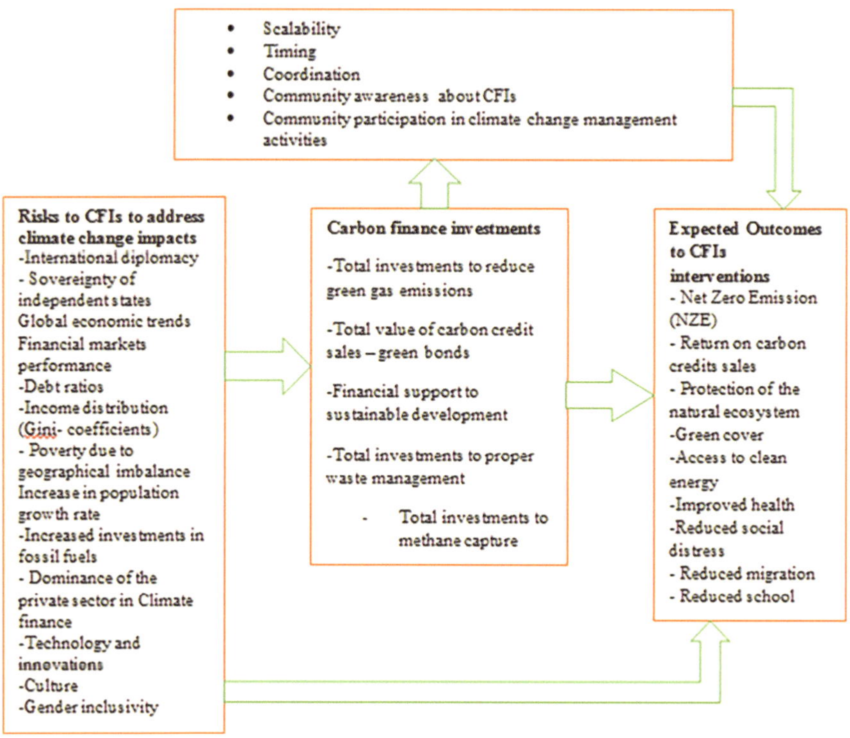

Figure 7.
The connectedness between CFIs, the risks of CFIs, and expected outcomes for CFIS to mitigate and control the impacts of climate change. Source: Designed by the researchers.

line with climate change [5]. The government plays a central role in directly supervising the work of corporations when it comes to issues like pollution plus monitoring the interventions at the different levels which are in line with mitigation, adaptation, and recovery from climate change impacts.

Governments can further intervene in the market mechanism to ensure macroeconomic stability by influencing market interest rates. These directly affect the cost of capital, the volume of CFI transactions, and the returns on CFIs. The governments' decision to reduce interest rates attracts more investors in the financial markets which can increase the scale of CFIs and vice versa is also true. Governments always find themselves with basically two competing alternatives which involve protecting the environment at the same time securing resources to promote economic growth [6]. However, through facilitating innovations such as the adoption of clean energy, governments can protect the environment while at the same time promoting economic growth and development [6].

The financial sector plays a vital role in providing the necessary infrastructure and instruments that can be traded among investors to promote climate change goals. Governments provide license, supervision, and advisory services to financial institutions to extend credit to those firms that embrace environmental and social responsibilities. Governments directly influence the business strategy of financial institutions to allocate resources toward environmental protection systems while at the same time promoting sustainable development [6] which helps to mitigate, adapt, and recover from climate change impacts.

2.3 Environmental risks of carbon finance investments policies on outcomes of climate change

Environmental risks involve those actions by humans that may give support or hinder the successful trading of CFIs to influence outcomes on climate change. These include; Agricultural practices which are in most cases for subsistence for the case of developing economies, and on a small scale. The increasing population increases pressure on existing resources and more so to the environment mainly through cutting down the green covers for cooking fuel [1]. The high population growth rate is further associated with the high expenditures to facilitate current consumption other than investments in technology and innovations by both governments and households which limit the available resources for CFIs.

CFIs are facilitated based on technology and innovations which require advanced use of Information Communication Technology (ICT) infrastructure with access to the internet in the management structures to ensure timely reporting and information flow. The issue of air pollution on financial decisions by corporations with pessimistic earnings forecasts is not taken seriously [19]. This is because such costs do not directly affect the operations of the business and companies as they opt to minimize such expenses on treatment of the fumes unless they are offered financial support through CFIs [20]. Another risk is when companies fail to disclose the true damage caused by their activities to the environment in a way of protecting their social equity (5) and trying to cover their regulatory requirements.

2.4 Social risks of carbon finance investments policies on climate change outcomes

Gender inclusivity has been identified as one of the major drivers toward achieving sustainable development goals. Including women is crucial for driving climate

DOI: http://dx.doi.org/10.5772/intechopen.1007154

change mitigation and adaptation as well as facilitating the transition to clean energy [11]. Many companies in EMDEs are now prioritizing climate action and advancing gender inclusiveness. These initiatives are still challenged by the ability to bring the two goals in an integrated, comprehensive, and strategic way in managerial positions to influence decision-making in climate change-related projects [11]. Gender inclusiveness necessitates the participation of women in top managerial positions for companies pursuing carbon finance initiatives as a means to address issues of climate change though with little success due to the dominance of men in managerial positions [11].

3. Strategies to minimize risk exposures associated with carbon finance investments

When causing harm to humans, climate change consequences neither adhere to specific laws or regulations nor do they take into account who is wealthier than who. Instead, there is no discrimination in the effects of climate change. This necessitates a number of non-discriminatory activities among the stakeholders in order to address the climate change challenges. Small and medium-sized businesses receive less funding in carbon finance despite being major contributors to greenhouse gas emissions and environmental harm. There is a need to find a mechanism by carbon finance investors relating to how the local communities can be empowered [4] with both resources and knowledge. This will promote innovations based on the available resources as compared to technological transfers. This should not be a "one size fits all" strategy, such as encouraging the use of renewable energy sources in rural areas, such as solar, hydropower, and natural gas power. Even while these actions are good at cutting greenhouse gas emissions and boosting investor profits, many African communities—many of whom are still poverty-stricken households—cannot afford them. This means that getting cooking fuel is always difficult. The scenario presents a serious risk to the ecosystem and the environment because natural trees are still being cut down massively for charcoal and firewood.

Secondly, the poor cannot participate in decision-making yet they are the ones hit most in terms of climate change impacts. There is a need to integrate the ideas and encourage low-developed countries to fully participate in CFI projects right from their planning to implementation. This will make communities own the innovations which can greatly reduce the monitoring costs and create benefits for all for sustainable development [7, 8]. The industrial strategy is a one-sided intervention yet it is the root cause of greenhouse emissions. There is a need to increase CFIs in other sectors such as agriculture and forestry which have significant contributions to climate change.

Governments should come up with tight supervision and regulation of the private sector especially investors in fossil fuels which accounts for the biggest percentage of greenhouse gas emissions. There is a need to increase the scope of renewable energy useto poor communities at very subsidized costs.

4. Conclusions

As opposed to technological transfers, empowering local communities based on available resources is probably going to yield more benefits when it comes to addressing climate change challenges. This can be through capacity building [4], political commitment, and creating more awareness about carbon finance investments in addressing climate change. Promoting local innovations that empower the local masses economically can help greatly to promote mitigation, adaptation, and recovery of climate change impacts. The increased involvement, participation, and awareness by households plus their ability to appreciate the environment through interventions such as CFIs can help to address the issues of climate change. The poor are less concerned about the environment but their focus is on day-to-day survival. Further, promoting local innovations can help to address the issues of poverty which in the long run creates a wide market for the new technology developed by the advanced nations to address climate change impacts.

As interventions for the mitigation, adaptation, and recovery of climate change impacts, CFIs benefit from a well-coordinated and regulated environment, which is made possible in part by international diplomacy. It accomplishes this by coordinating resources and making sure projects are closely watched to make sure their intended objectives are met. Low- and middle-income nations do, however, participate in the planning processes to a limited extent. These nations suffer immensely from the effects of climate change and are frequently the implementing agencies. As a result, their capacity to provide ideas and information for developing suitable techniques to handle climate change challenges is limited.

Increasing stakeholder participation mainly in the early stage for CFIs promotes the achievements of CFIs goals. Stakeholders should essentially be included in the planning stages of projects in order to gather the best ideas and learn how to best manage the risks associated with climate change and make use of its opportunities through CFIs. There is a need to empower communities with resources to engage in productive activities and improve their socio-economic status in order to appreciate the environment.

Acknowledgements

First of all, our sincere appreciation goes to all the sponsors of this book chapter.

Secondly, Professor Israel Kibirige for his constructive criticism and guidance while writing this book chapter.

Conflict of interest

The authors declare no conflict of interest.

Notes

Dr. Nabulya Jesca's ORCID number: 0009-0008-4196-9332.

Author details

Nabulya Jesca[1]*, Peter Oseghale Ohue[2] and Jerome Ntege[3]

1 Bugema University, Kampala, Uganda

2 University of Ibadan, Nigeria

3 Kyambogo University, Kampala, Uganda

*Address all correspondence to: jescanabulya@gmail.com;
jnabulya@bugemauniv.ac.ug

References

[1] Dinh HBP, Vuong TT, Tee CM, Anh L. Carbon risk and corporate investment: A cross—Country evidence. Elsevier Journal. 2021;**46**:1-13. DOI: 10.1016/j.frl.2021.102376

[2] Trotsenburg AV. New WBG Academy: Building a Cadre of Development and Practioners. World Bank; 2024. Available from: https://blogs.worldbank.org/en/voices/new-wbg-academy-building-a-cadre-of-development-leaders-andpractitioners?cid=ECR_E_NewsletterWeekly_EN_EXT&deliveryName=DM223258

[3] Nobanee H, Dilshad MN, Abu LO, Ballool B, Al Dhaheri S, Almheeiri N, et al. Insurance for climate change and environmental risk a bibliometric review. International Journal of Climate Change Strategies and Management. 2022;**14**(5):440-461. DOI: 10.1108/IJCCSM-08-2021-0097

[4] Bosi M. The World's Ten Years of Experience in Carbon Finance: Insights for Working with Carbon Markets for Development and Global Greenhouse Gas Mitigation. 2009. Available from: https://unfccc.int/resource/webcast/collections/091102_AWG_Barcelona/downloads/091105_Martina_Bosi.pdf

[5] Fu C, Lu L, Pirabi M. Advancing green finance: A review of climate change and decarbonization. 2024;**2**:1-23. DOI: 10.1007/s44265-023-00026-x

[6] Wu G, Liu X, Cai Y. The impact of green finance on carbon emission efficiency. 2024;**10**(1). DOI: 10.1016/j.heliyon.2023.e23803

[7] Ummar R, Sitara K, Cheong CWH. From Turbulence to Resilience: A Bibliometric Insight into the Complex Interactions between Energy Price Volatility and Green Finance; 2024. DOI: 10.1016/j.energy.2024.131992. Available from: https://www.researchgate.net/publication/381378235_From_Turbulence_to_Resilience_A_Bibliometric_Insight_into_the_Complex_Interactions_Between_Energy_Price_Volatility_and_Green_Finance

[8] Climate Policy Initiative. Global Landscape of Climate Finance 2022. San Francisco, CA: Climate Policy Initiative; 2023. Available from: https://www.climatepolicyinitiative.org/publication/global-landscape-of-climate-finance-2023/

[9] LSE. What is Climate Change. 2023. Available from: https://www.lse.ac.uk/granthaminstitute/explainers/what-is-climate-finance-and-where-will-it-come-from/

[10] Rusydiana AS. Bibliometric Analysis of Scopus-Indexed Climate Finance Studies. 2024. DOI: 10.58968/tae.v1i1.451

[11] Gabala NA, Nyinkindi E. A Road for a Greener, Gender-Inclusive Future. 2024. Available from: https://blogs.worldbank.org/en/voices/a-roadmap-for-a-greener-gender-inclusive-future?cid=ECR_E_NewsletterWeekly_EN_EXT&deliveryName=DM225200

[12] IEA. World Energy Investments 2023. 2023. Available from: https://www.iea.org/reports/world-energy-investment-2023/overview-and-key-findings

[13] United Nations Framework Convention on Climate Change. Climate Finance and Carbon Markets. Bonn,

Germany: UNFCCC; 2023. Available from: https://unfccc.int/sb58

[14] World Bank. The Role of Carbon Markets in Climate Change Mitigation. Washington, DC: World Bank; 2022. Available from: https://www.worldbank.org/en/topic/climatechange/brief/the-world-bank-engagement-roadmap-for-carbon-markets

[15] London School of Economics. Financing Climate Action: The Role of Carbon Finance Investments. London, UK: London School of Economics; 2023. Available from: https://www.getsmarter.com/products/universities/the-london-school-of-economics-and-political-science

[16] Climate Policy Initiatives 2021. Global Landscape of Climate Finance. 2021. pp. 1-48. Available from: https://www.climatepolicyinitiative.org/wp-content/uploads/2021/10/Full-report-Global-Landscape-of-Climate-Finance-2021.pdf

[17] International Energy Agency. World Energy Investment 2022. Paris: IEA; 2022. Available from: https://www.iea.org/reports/world-energy-investment-2022

[18] Wambui R. Financing Climate Action: Aligning Financial Flows with Global Goals. Grantham Research Institute on Climate Change and the Environment; 2023. Available from: https://www.climatepolicyinitiative.org/wp-content/uploads/2023/11/Global-Landscape-of-Climate-Finance-2023.pdf

[19] International Energy Agency. Net Zero by 2050: A Roadmap for the Global Energy Sector. Paris, France: International Energy Agency; 2023. Available from: https://www.iea.org/reports/net-zero-roadmap-a-global-pathway-to-keep-the-15-0c-goal-in-reach

[20] Jia Q. The impact of green finance on the level of decarbonization of the economies an assessment of the United States, China and Russia's current agenda. Business Strategy and the Environment. 2023;**32**(1):110-119

Section 2

Nature's Role and Technological Solutions

Chapter 3

Blue Carbon: The Underwater and Coastal Alleate of our Uncertain Future

Martina Gaglioti

Abstract

Blue carbon is the first ally to face a changing climate. Despite its importance, so many people are not enough aware of the importance of these natural defenses' preservation. Also, there are still so many knowledge gaps in how we can perform a coordinated plan of restoration interventions or meaningful ocean prediction trends. Therefore, the blue education is a precious support in shaping the future of our ocean-related knowledge even in terms of nature-based solution development and new promising strategies to cope even with the most demanding future scenarios. Ranging from the alarming rise in seawater temperatures to the acidification of our oceans, all these phenomena require an integrated approach to be meaningfully addressed. The ocean-climate nexus is among the most significant challenges outlined within the UN Ocean Decade and is among the main urgencies to be addressed through specific EU policies. A healthy ocean is synonym of healthy people: aspects inextricably linked in the "One Health" concept, meaning a strong convergence between human, animal and environmental health. Beside the climate-related issues affecting the ecosystems, heatwaves and soil consumption fuel the spread of vector-borne syndromes or zoonoses, representing to date 60% of emerging infectious diseases globally. Indeed, learning from nature is an asset to raise our comprehensive preparedness capability.

Keywords: blue carbon, ocean literacy, remote sensing, field engagement, restoration, blue education, one health

1. Introduction

Blue carbon is defined as the carbon captured and stored by the oceans. Usually, when we refer to blue carbon ecosystems we are referring to: seagrass meadows, tidal marshes or mangroves forests [1–3]. These three ecosystems occupy less than 0.2% of the seafloor and alone are able to sequester more than half of the ocean's carbon. The ocean takes up about 25% of the carbon we add to the atmosphere, slowing climate change effect, reducing its intensity and giving us more time to mitigate and adapt to its consequences. Providing healthy marine ecosystems, we are acting to preserve these precious allies against the climate crisis, providing at the same time wider environmental and human-wellbeing benefits. Imagining a journey developed from the sea surface toward the deepest part of the Ocean, blue carbon in all its variegated forms includes the following.

1.1 Carbon absorbed by aquatic plants, algae and phytoplankton

The photosynthetic activity enacted by marine plants and algae can be considered as a golden climate mitigation opportunity since they are doing a lot of the heavy lifting in terms of climate change mitigation, with an efficiency rate higher than 40–50 times than the tropical rainforests [4].

But our underwater allies are not only these "rooted characters," other unneglectable actors of the climate challenge are the phytoplankton species, supplying at least half of the world's oxygen, absorbing immense quantities of carbon locking it away from the ocean when they die or becoming part of the marine food webs.

1.2 Carbon stored in the bodies of living animals

This is the case of the marine animal forests [5] dominated by gorgonians and hard skeleton corals or the millennial coral skeletons, such as the fossilized fragments coated from ferromanganese structures, lying on the American Samoa seafloor and detected during the telepresence connection onboard E/V Nautilus during the NA165 expedition in the E Mamana Ou Gataifale II sites (**Figure 1**).

1.3 Carbon sequestered in deep sea sediments

The ocean seafloor hosts the largest pool of sediment/soil carbon stocks of the planet and are an important part of the global carbon cycle [6]. The current storage of marine sediments is estimated to be 2322 Gt of carbon in the top meter of sediments. Around 79% of ocean sediment carbon is estimated to be stored in abyssal plains or basin zones, such as the Clarion-Clipperton zone [7, 8]. This carbon can be periodically resuspended through human-driven disturbance, even in deep sea areas, where bottom-trawling activities are performed or due to mechanical disturbance due to deep sea cables and infrastructure development causing significant adverse impacts [8]. Sediment plumes deriving from the action of machines used for deep sea mining [9]

Figure 1.
Marine animal forests and hard skeleton corals detected during E/V Nautilus expedition seasons 2023–2024.

could have direct consequences on the carbon sequestration rate taking place in ocean waters, and the sediment resuspension may result in an higher level of turbidity, blocking sunlight and reducing photosynthetic available radiation used for primary production process. This impact is impaired with another meaningful source of pollution, which is the acoustic pollution deriving from machineries acted noise. Considering that many polymetallic nodules are covered by plants and primary producers, such an impactful practice can directly affect also the carbon sequestration processes carried out by these living components (**Figure 2**). These impacts have not yet been precisely calculated, but considering that nodules take tens of millions of years to shape, many of these damages would essentially be considered as permanent (**Figure 3**).

1.4 The ocean carbon pumps

The chemical equilibrium reaction occurring in surface waters determines the release of dissolved carbon into three chemical forms: carbon dioxide, hydrogen carbonate and carbonate. These chemical species concur at the marine carbon cycle which can be acted in three different ways, all defined as "carbon cycle pumps":

1.4.1 The physical carbon pump

This physical component of the carbon pump is driven by the ocean currents and the differences occurring in temperature and salinity. When its equilibrium is altered or disrupted as a consequence, we can have an alteration of the thermohaline circulation (**Figure 4**). The physical carbon pump can distribute the dissolved carbon (carbon dioxide, hydrogen carbonate and carbonate) through the water masses moving toward the shallower ocean layers or sinking toward the deeper waters.

The physical carbon pump is the first mechanism for the transport of carbon related to anthropogenic emissions. This process is particularly relevant in the polar regions due to the higher solubility of CO_2 in colder waters. Subsequently, the dense and cold waters originated at pole level are spread over the globe thanks to the global conveyor belt, driving the thermohaline circulation at planetary scale.

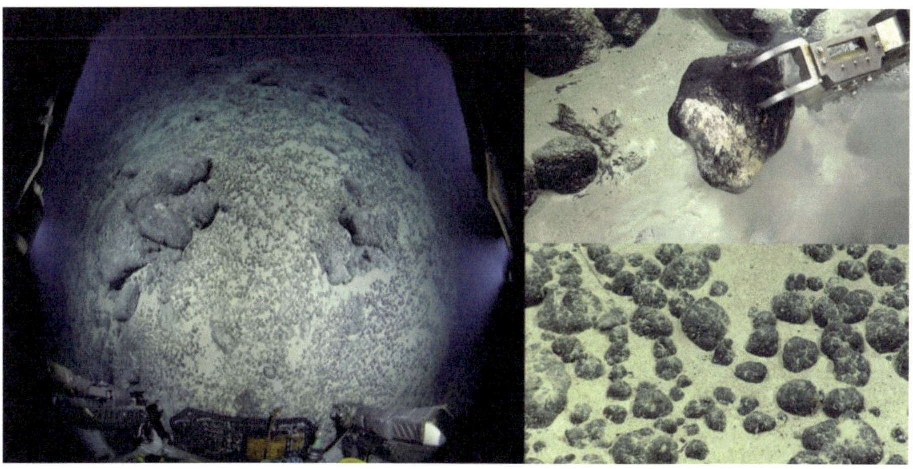

Figure 2.
Polymetallic nodules detected during a deep exploration by the ROV during the NA166 expedition off American Samoa.

Figure 3.
Some exemplificative pictures showing some human-driven impacts, mainly associated with lost fishing gears and land-sourced debris, detected during the exploration surveys of the NA154 expedition (Source: E/V Nautilus expedition 2023—Papahānaumokuākea Marine National Monument).

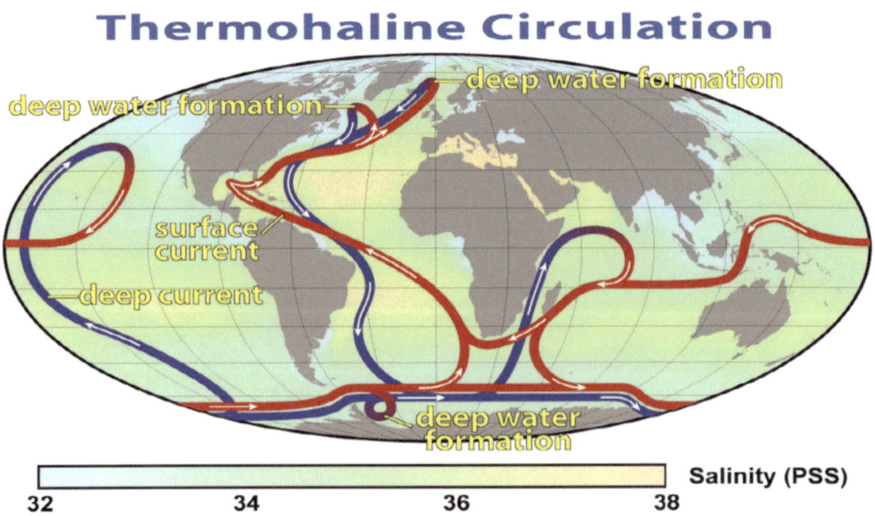

Figure 4.
Thermohaline circulation and the conveyor belt mechanism driving the water masses at global scale (Image source: NASA Earth Observatory).

DOI: http://dx.doi.org/10.5772/intechopen.1008533

1.4.2 The organic biological carbon pump

The organic component of biological carbon pump is driven by the biological communities living in the surface waters of the ocean. This is where photosynthesis is carried out by vegetal organisms using solar energy to fix inorganic carbon and produce biomass. When these primary producers are eaten, the consumers ingest also the carbon included in their bodies, and this is turned into muscular mass or body fats and in the end excreted as fecal pellets. The organic matter natural processes lead this carbon through the entire food web. A part of the organic matter when not consumed or biologically transformed in the surface layers of the ocean waters (in this case it is called the "recalcitrant organic matter") can reach the seafloor and encounter micro-organisms entering into the so-called microbial loop. Through this mechanism the stored carbon is again released into the seawater in the form of CO_2. The remaining material settles into the deep sea as "marine snow."

1.4.3 The inorganic biological carbon pump

Beside the photosynthetic processes, there is a second mechanisms allowing the marine organisms to biologically fix the dissolved carbon from the water and then transport it to deeper zones of the Ocean. This is the achieved through the construction of calcareous shells or skeletons. This process, in the scenario of climate change, is at high risk due to the ocean acidification issue, which may accelerate the rate of carbonate dissolution in the warmed and more acidic waters predicted even from the most optimistic RCP scenarios. The main categories of marine organisms that may be negatively affected from these circumstances are bivalves, corals, crustose coralline algae, foraminifera. Once they die their skeletons can sink in the seafloor where they can be incorporated into the sediments (**Figure 5**).

At EU level, there is a high level of attention around the topic of biological carbon pump and recently, also an EU Horizon Project financed specific research on these processes within the research initiative named *OceanICU* (**Figure 6**). This project will last for 5 years and has among its main goals leading to a better approach to

Figure 5.
A Nautilus shell buried in the seafloor sediment, detected during the expedition NA165 in American Samoa in summer 2024.

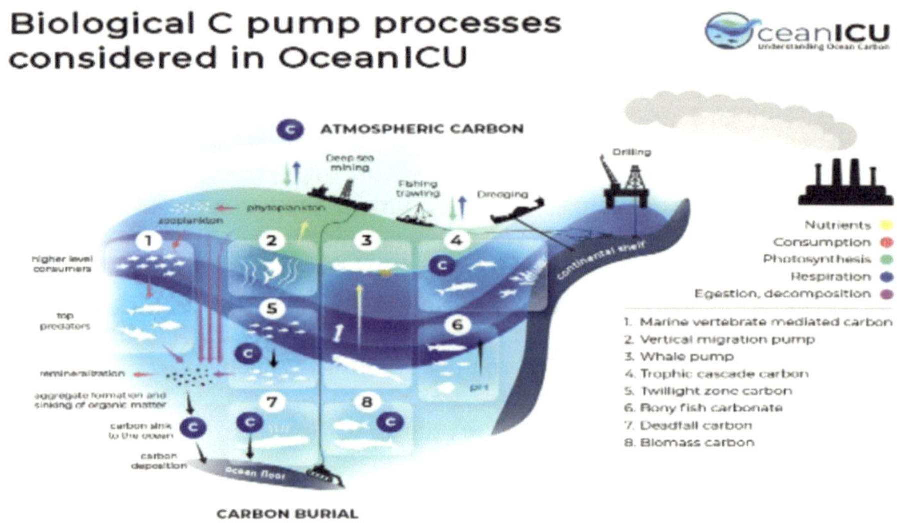

Figure 6.
A diagram on biological carbon pump processes extracted by the OceanICU Horizon Europe Project infographics (Source: https://ocean-icu.eu/).

understand the climate change and align the future endeavors with the EU Green Deal reaching the ambitious targets of Zero by 2050.

1.5 The Ocean State Report through remote sensing: An eye on earth through CMEMS data

The ocean is a major heat sink, absorbing about 90% of the excess heat trapped in the Earth's climate system. Since 2005, the pace of ocean warming has almost doubled. The Copernicus Marine Service Ocean State Report is a milestone publication providing a state-of-the art assessment of the current state of the Global Ocean and the EU regional seas both in a scientific perspective and with the aim to support the decision-making processes and policies. In this report, a multi scale 4-D approach has been developed, including information from remote-sensed data, *in situ* measurements, complex analyses performance and multi-level approach to the main components of the marine environments: the blue dimension (considering oceanographic parameters such as: currents, hydrographic conditions and tides), the white one (including the ice-coverage estimation and the ice-cores analysis for backward estimation on paleoclimatic conditions) and the green one (related to the primary production rates including the photosynthetic processes) [10].

According to the Ocean State Report 8 (OSR8) in August 2022, the coastal waters surrounding the Balearic Islands recorded the highest sea surface temperature in 40 years! Reaching 29.2°C, this was over 3°C higher than the average for the past few decades. The latest report reveals that European regional seas have been affected by more intense, frequent and prolonged marine heatwaves [10]. Namely, marine heat waves (MHW) are defined as heating episodes in which the temperature is significantly higher than normal for a minimum of five consecutive days (**Figure** 7) [11]. These events can occur in many edges of the ocean and their magnitude and frequency have increased over the last decades, with harmful consequences and impacts on ecosystems and human livelihoods. According to the data recently published Ocean

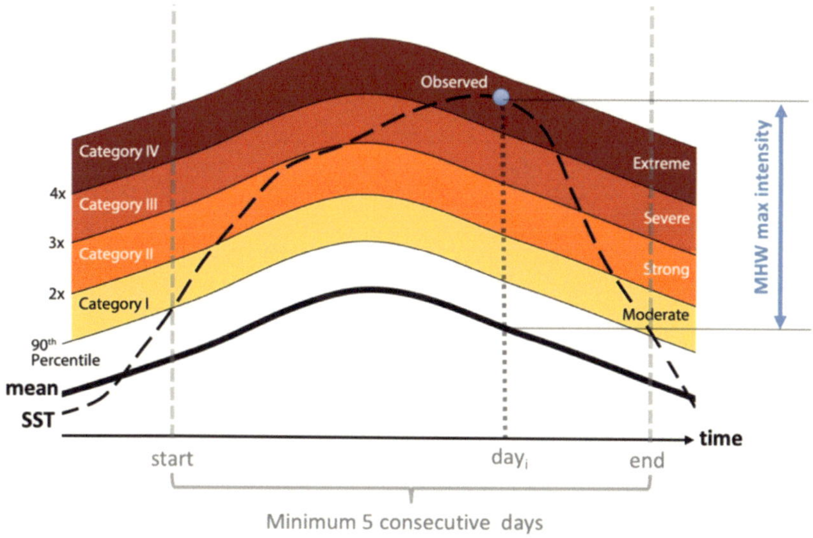

Figure 7.
A diagram explaining the threshold level for marine heatwaves definition (Image source: Ocean State Report 8).

State Report 8 [12], the Mediterranean Sea is one of the sea basins most affected by these extreme phenomena both due to the high anthropogenic pressures acting on this small-scale basin and due to the oceanographic conditions making it extremely vulnerable to abrupt variations and extreme phenomena rate intensification.

2. Marine biodiversity in a warming global ocean

The most tangible consequence of extensive marine heat waves is the enhancement of mass mortality event occurrence directly affecting an increasing number of species and habitats. In particular, benthic organisms, such as corals, coralline crustose algae and sponges, are particularly prone to extreme thermal stress since unable to actively swim to move toward deeper and cooler zones when these temperature rises extensively occur. This may result also in disruption in fish ecosystems and halieutic resource depletion since some fish species tend to migrate seeking more temperate waters changing their usual migration pathways, and more thermophile species from other regions (often invasive) replace the native ones [13].

2.1 Marine heatwaves more frequent than unexpected events

The massive coral bleaching events are reversible phenomena. According to the most relevant expectations foreseen even by the latest projections expressed in terms of Representative Concentration Pathways (RCPs) by the Intergovernmental Panel for Climate Change (IPCC) in a range from low (RCP2.6) to high (RCP8.5) greenhouse gas emission scenarios, within 2050 corals will be extinct (**Figure 8**) [14, 15].

Massive bleaching events are due to the disruption within the balanced relationship between corals and their photosynthetic symbionts (zooxanthellae algae) with the consequent loss of photosynthetic pigments from these algae, causing coral tissue to become translucent and making underlying white skeleton visible. Beside this

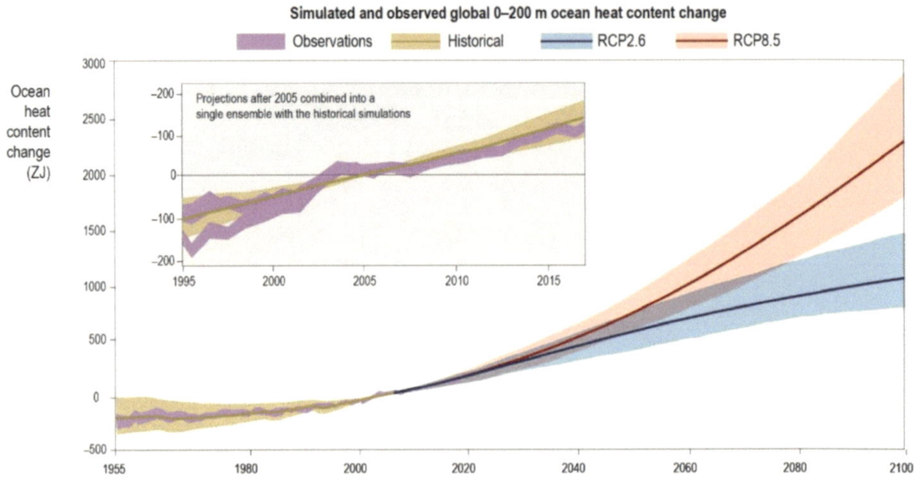

Figure 8.
Simulated, observed and projected ocean heat content from 1995 to 2100. Projections are simulated for two representative concentration pathways (RCP), RCP2.6 (blue) and RCP8.5 (red). Source: Bindoff et al. [11].

well-known phenomenon, the occurrence of infectious diseases such as white plague or the back band disease is commonly observed in tropical reef assemblages [10, 14, 15].

The consequences of these ecosystems shift may be dramatically considering the role that coral reefs act as biological treasures of the ocean despite being daily threatened by a myriad of stressors. According to a recent assessment conducted applying the IUCN Red List assessment criteria [16] to 845 zooxanthellate reef-building coral species to date one-third of reef-building corals face elevated extinction risk from climate change and local impacts and 32.8% are in categories with the elevated risk of extinction [15]. Coral reefs can be considered as one of the marine ecosystems with higher densities of biological diversity, deeply interconnected in terms of ecological interactions (competition, symbiosis, predation…), but at the same time highly vulnerable to abrupt regime shift consequences. Globally, at least 350 million people and a quarter of all marine life depend on coral reefs for their sustenance and livelihood [13]. Fostering the establishment of a cohesive "Coral Disease Research" community, develop a shared guidance for the proper handling and containment of corals in infectious disease experiments, create and support advanced educational opportunities on these topics while developing a systematic approach to investigate and study diseases in corals, provide competitive-based grant opportunities to fill knowledge gaps, and adopt model coral species for research to better characterize basic knowledge on physiology and ecology of corals in healthy conditions to better address management plans useful to reduce stressors that may make corals more vulnerable to disease. Develop a systematic approach to investigate and study diseases in corals to make responsiveness to epidemiology more impactful, adopting an ecological epidemiology approach to identify risk factors and assess their contribution to coral reef degradation in a timely and effective manner [14].

According to the Global Distribution of Coral Diseases from the Reefs at Risk report [15], the anomalies affecting these structuring ecosystems over the last few decades were alarmingly distributed worldwide (**Figure 9**). Similarly, mass mortality events are commonly detected in temperate regions, such as the Med Sea basin (**Figure 10**), where the coralligenous assemblages including hard-skeleton corals and sponges are similarly affected by prolonged heath waves.

DOI: http://dx.doi.org/10.5772/intechopen.1008533

Figure 9.
Some examples of coral diseases and reef degradation detected during scientific surveys off Bangka Island (North Sulawesi, Indonesia) in October 2019.

Figure 10.
Some examples of heat-waves effects detected in bleached coral colonies from Pacific Ocean shallow waters detected off Malapascua Island coastal waters over the time period: 2018–2024.

Recently published data on risk scenarios for coastal ecosystems based on projected climate impacts show how there are so many blue carbon storing ecosystems, which may be negatively affected by the intensification of warming and climate hazard-related scenarios. These threats beside ocean warming include acidification, deoxygenation, nutrients and particulate organic carbon flux shifts [12].

There is a metric called Degree Heating Weeks (DHWs) developed to help coral reef managers worldwide monitor bleaching risk. This is used by NOAA's Coral Reef Watch program [14] to predict and monitoring at large scale the occurrence of mass bleaching events due to the combined effect of prolonged excess of high sea temperatures exceeding the normal summer maximum and the entity of the sea temperature increase causing the heat stress on corals and other marine ecosystems. A threshold value of 2DHWs is the signal for the likely onset of a global bleaching event (**Figure 11**).

A recent field expedition performed through SCUBA diving sessions in the Pacific Ocean off Malapascua Island highlighted how coral bleaching events are not uncommon even in this area. Despite an overall good health status of the reef-building corals I came through some examples of pleached branches or early-stages coral bleaching phenomena (**Figure 12**).

2.2 Global sea-level rise and the two main causing factors: Thermal expansion and increased melting of ice sheets

The IPCC reports that global sea-level rise has accelerated to 3.25 mm per year since 1993. By 2100, sea levels are expected to rise by 430 to 840 mm [17]. There are some discrepancies on the fact that SLR will have only negligible impacts on coral reefs. Some studies prediction a rise rate of 2–3 mm per year considering that due to coral growth and reef accretion rates the reefs can keep pace with these variations. Other studies underlay how due to the multiple stressors acting on corals and other marine ecosystems the overall accretion of reefs may be much slower than the gross rate of individual colonies, predicting even under RCP 4.5 the inability of the majority of reefs to compensate the SLR rate. Also, the phenomenon will deeply affect the current asset of coastal zones and human settlements with tangible socioeconomic consequences affecting human livelihoods, infrastructures, food security and climate mitigation at the coast and beyond.

The issue will be tangible even in terms of ecosystem service provisioning: When sea-level rise occurs, reefs can end in deeper water, and the strength of some of the

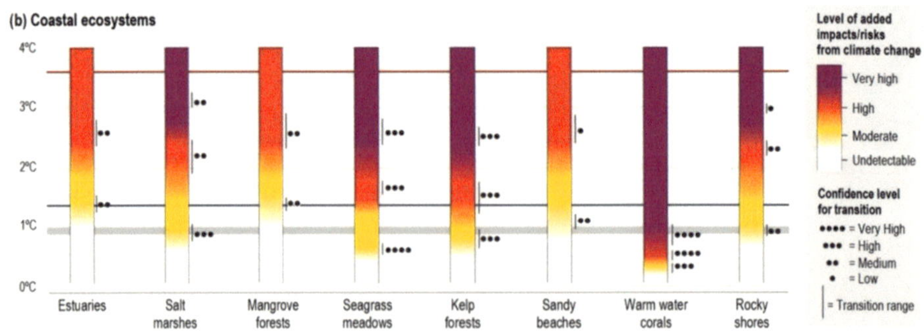

Figure 11.
Projected risk scenarios for coastal ecosystems, which may be affected by climate impacts. Present day (gray line) corresponds to the 2000s, whereas the different greenhouse emissions scenarios, RCP2.6 (blue line) and RCP8.5 (red line), correspond to 2100.

Figure 12.
A massive hard-skeleton coral (Acropora spp.) in a early-stage bleaching phase (Malapascua Island MPA, Photo credits: Martina Gaglioti).

ecosystem services they provide to coastal communities may become weaker, particularly in terms of mitigating wave action, reducing coastal erosion, and protecting shorelines during tropical storms. Sea-level rise can also inundate and erode coastal habitats such as mangroves and turtle nesting beaches, causing an increased sedimentation due to shoreline erosion, which could smother reefs or reduce the photosynthetic available radiation (PAR) incidence [18]. Similarly, saltwater intrusion (SWI) intended as the intrusion of saltwater into freshwater resources will affect both surface waters and groundwater. It will mean severe risks for agriculture, freshwater availability and coastal communities' livelihoods as a consequence of salt damage on crops and health risks associated with salination of drinking water. SWI reduces freshwater storage and impacts soil fertility, vegetation, freshwater species and ecosystem services, especially in transitional ecosystems characterizing deltas and estuarine regions [17, 19].

Even in the Philippines area explored at the end of 2024 and early January 2025, local stakeholders highlighted similar issues due to unmanaged wastewaters in coastal areas and tangible plastic pollution cases due to widespread presence of unrecycled wastes. Due to the proximity of coastal communities to the seashores and their extensive dependency from the ocean for livelihoods and wellbeing, these gaps in waste management and the lack of awareness on the human-driven impact may be even more dramatic in terms of consequences. For this reason, an extensive dissemination and outreach effort engaging locals would be more than desirable (**Figure 13a** and **b**).

2.3 The role of whales in ocean carbon cycling

Did you know whales can act a pivotal role in mitigating climate change impacts by storing carbon in their bodies and transporting nutrients that benefit ocean food

Figure 13.
Seashores affected by plastic pollution and marine litter: (a) locals from Malapascua island daily engaged in the marine litter collection; (b) some debris on the shores of Maya Harbor in the nearby of the coastal settlement.

chains? About 31% of carbon dioxide emissions are captured by the ocean, removing carbon from the atmosphere that would otherwise continue to trap heat and increase temperatures [20].

Scientists believe whales contribute to all three of the known carbon storage mechanisms. Targeted conservation plans for these charismatic species may help in mitigating climate change by increasing blue carbon capture, benefiting marine and terrestrial *species alike [Pearson et al].* They are likely to be responsible for the support of even greater amounts of blue carbon storage before their populations were dramatically depleted by commercial whaling in the 1800s [21].

Like trees in a rainforest or marine algae in a kelp forest, whales are efficient carbon sinks capturing and storing atmospheric carbon directly in their large bodies, being also quite longeval animals. When whales die, their carbon-rich carcasses often sink to the seafloor creating the so-called whale fall ecosystems around their bones and flash remains and at the same time, a large amount of that carbon they stored throughout their life is somehow prevented to return the atmosphere as CO_2. Within the so-called "whale falls" during their decomposition whales also indirectly contribute to carbon capture by providing nutrient-rich waste to phytoplankton that use it for primary production, while absorbing large amounts of carbon dioxide.

Also, as testified from the images collected over time during the deep sea expeditions near the Clayoquot Slope, the skull and vertebrae of the whale skeleton host a rich community of benthic fauna that adapt to make their home on top of it. This site was discovered for the first time in 2009 by an expedition led by the Monterey Bay Aquarium Research Institute.

The "boney buffet" around the whale carcass provides an abundance of nutrients for organisms living near the seafloor (**Figure 14**). Through photogrammetry techniques, it is possible to record high-resolution videos of the entire whale skeleton. Photogrammetry video surveys provide also data on decomposition rates, helping us understanding how whale falls support deep sea ecosystems and at which rate the organic matter transformation processes occur [22–24].

Another interesting aspect is the role migrating whales act in nutrient cycling across long distances. Indeed, in addition to their vertical movements they do through the water column, most whale species migrate seasonally from nutrient-rich feeding grounds to nutrient-poor breeding grounds to mate and give birth. The nutrients

DOI: http://dx.doi.org/10.5772/intechopen.1008533

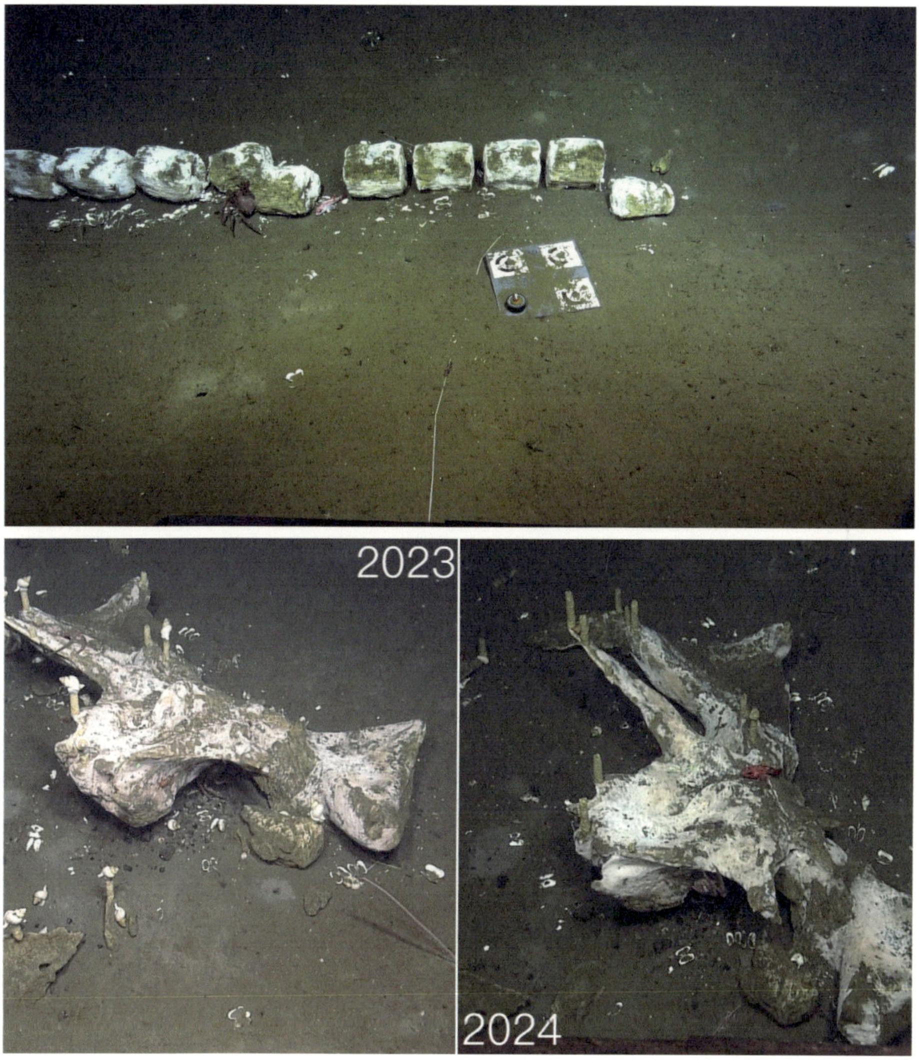

Figure 14.
Whale fall ecosystems associated with whale bones in Canadian waters in a monitoring site discovered in 2009 during the NA161 ONCabyss expedition with Ocean Network Canada near the Clayoquot slope.

whales consume on their feeding grounds are expelled as feces and urine along their migratory routes and in their breeding grounds. This stimulates phytoplankton blooms and increases carbon capture *via* photosynthesis. Baleen whales embark on some of the longest migrations on the planet—up to 12,000 miles. These large whales transport nutrients across oceans and encourage phytoplankton blooms along their migration corridors.

Being native of the first designated Biosphere Reserve, I had also the honor to encounter some specimens of cetacean over my life pathway. Pontine Islands and Campanian coast are both currently designated by IUCN as "Important Marine Mammals Areas" (IMMAs) due to their importance for monitoring and conservation of marine mammals. The Campanian and Pontine Archipelagos is an important habitat for Endangered Mediterranean sperm whale (*Physeter macrocephalus*). Groups of

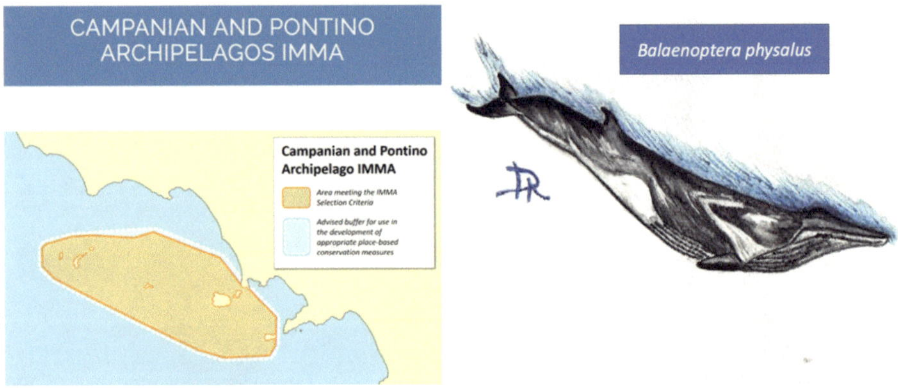

Figure 15.
The extension of the IMMA designated between Pontine waters and Campanian coasts and the drawing of the E. robustus *specimen made from a local artist in 2021.*

solitary males, loose male aggregations, bachelor groups, and social units of females with immature whales have been observed to coexist in the area. Approximately 60% of the identified individuals showed site fidelity. This area has also been identified as a feeding site, the complex topography of canyon systems being a driving factor shaping the species' presence and distribution [25]. One of the most recent and surprising sights of a gray whale (*Eschrichtius robustus*) specimen occurred in Ponza waters on April 2021 and a *Balaenoptera physalus* specimen recorded in the east coast of Ponza in October 2020 (**Figure 15**). Both the occurrences were signaled by island native people and confirmed by local experts, underlining the importance of collaborative approaches for timely phenomena detections. In general terms, marine mammal stocks are expected to respond to climate change and shifts in distribution, abundance and/or phenology may be observed over time even if some aspects of these population dynamics change still understudied. Vulnerability assessments can help to identify candidate stocks for targeted climate-related research and targeted conservation measures [26].

2.4 Climate refugia for VMEs

Deep sea environment are not exempt from climate crisis effects. Pressures related to climate change are likely to influence benthic ecosystems in general, and vulnerable marine ecosystems (VMEs), in particular, through direct or indirect mechanisms. VMEs are recognized as having high ecological significance and susceptibility to disturbances, including climate change. There are some indicator taxa, which may be particularly significant to identify climate refugia under future climate change scenarios. The best way to provide information about these ecosystems especially in remote areas such as the seabed of areas beyond national jurisdiction (ABNJ) and deep sea environment is implementing spatial predictions for VME indicator taxa. These findings on cumulative effects detectable on VMEs are also relevant to better address future marine conservation policies and implement relevant ecosystem management and spatial planning strategies [27]. According to the outcomes of the latest ICES Report on VMEs and significant adverse impacts on these areas, it is important to underline clearer management measures and uniquely identify indicator taxa for better informed conservation measurement implementation, under the upcoming climate scenarios [28].

DOI: http://dx.doi.org/10.5772/intechopen.1008533

3. The climate-related challenges within the UN ocean decade and beyond

3.1 The more than ever-needed efforts toward the ocean-climate nexus

At the basis of the ocean-climate nexus, there is the strict interconnectedness between the forces acting on climate processes at planetary scale and the biological processes we are engaged in along with all the other organisms living on Earth. The ocean has a pivotal role in maintaining this equilibrium and sustaining life. Its role is also essential in mitigating climate change acting as the major carbon sink and absorbing heat. Despite its role as buffer for CO_2 emissions, its potential to counteract the emissions related to human activities is decreasing and we are experiencing some negative effects such as the acidification processes acting on the marine food web as a whole affecting directly carbonate-shelled organisms and vertebrate skeletons. The crucial role marine ecosystems are acting in climate regulation that is a priority to be addressed through educational programs and dissemination initiatives and it is also a political priority at EU and global level.

3.2 Tackling climate crisis while sustainably nourishing global population

Climate change threatens our skills to ensure global food security. Climate change has both direct and indirect impacts on agri-food systems due to shifting and unpredictable rainfall patterns and temperatures, a higher incidence of extreme weather events and disasters such as drought, flash floods, outbreaks of pests and disease and ocean acidification. According to the United Nations Framework Convention on Climate Change (UNFCCC), FAO is supporting countries in the adoption of adaptation plans to adapt and tackle with climate crisis and to mitigate climate change by reducing or preventing greenhouse gas emissions, through its projects and programs and a wide range of knowledge products [28, 29]. To meet the strategic ambition of the Ocean Decade Challenges, the "Ambition, Action Impact" report [16] outlines a set of high-level priorities for ocean science and knowledge over the coming years that look at the interactions between the challenges and the enabling conditions essential to ensure the development of relevant and timely scientific knowledge useful to support the decision-making processes. These include tackling marine pollution, advancing ecosystem-based management, exploring deep sea biodiversity, encouraging climate-resilient projects and finding valuable solutions to reduce the impact of fisheries on marine biodiversity sustainably nourishing the global population. Both in the ocean and on land, the correct resource management is crucial to reduce the long-term impacts on natural resources and consequently on ecosystems' functioning [30]. To do that, we need to rethink our food system enabling a multistakeholder approach even reshaping the extant rules related to the fisheries resource management [31].

4. The fascination of ocean exploration in a warming world

4.1 Insights from field expeditions

Nowadays, technology not only allow us to rely on a huge network of experts connected globally to join forces and expertise seeking for planetary crisis solutions, but we are able to cross the respective national boundaries enabling explorations even hundreds of meters below the sea surface, through telepresence technology. This is what happened for instance in 2023 and 2024 onboard E/V Nautilus during the exploration campaigns operated by NOAA and Ocean Exploration Trust or during the Arctic exploration campaign in the Loki's castle hydrothermal vent field, held by the scientists from the Center for Deep Sea Research of Bergen University (Norway). In the first case, the deep sea mapping expedition across the Pacific Ocean occurred engaging scientists and science communicators from all over the World and the cruise plans were planned engaging in the process and interacting with the indigenous people to make their knowledge an integral part of the expedition planning and scientific outcomes. The second initiative allowed the remotely operated connection with home-based scientists during the deep sea exploration of an iconic venting site located at 2300 m below the surface. The relevance of this approach resides in the fact that deep sea still being largely unknown and understudied being inaccessible for most of the people and being really expensive to study it. Thanks to livestreaming and telepresence technology this gap is going to be partially filled and also, the knowledge advancements and the knowledge gaps fillings are going ahead. The images below for example come from a ROV dive conducted onboard the research vessel G.O. SARS within the RIALHE project (Remote Intelligent Access to Labs in Higher Education) in collaboration with the Centre for Deep Sea Research at University of Bergen with the main objective to target students, educators, researchers and general public. In this case a deep sea hydrothermal vent community is highlighted and these studies are really important even in the climate-related investigation aimed at the understanding of inorganic carbon fixation mechanisms, macrofauna distribution and vent-obligate and background non-vent taxa distribution in the seafloor space (**Figure 16**) [32].

4.2 CO_2 emissions and venting sites as natural laboratories for climate scenario predictions

For decades, the scientific community used the marine CO_2 emissions as valuably proxies for climate change scenario investigations. They usually occur in volcanic active areas both in coastal areas and in deep sea environments.

Hydrothermal systems in Panarea [22] have been used also as training places for scientific diving: in particular the shallow venting site located off Bottaro islet in Panarea (Eolian Islands). In these natural emission fields, many investigations on benthic communities and diversity gradients occurred over the years (**Figure 17**) and the main findings

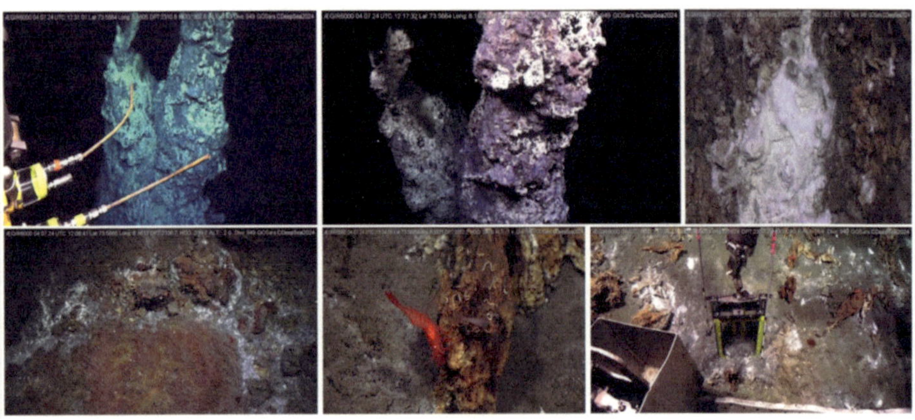

Figure 16.
Hydrothermal emissions and deep sea vents in the Arctic region (2024_Loki's castle hydrothermal field).

DOI: http://dx.doi.org/10.5772/intechopen.1008533

revealed meaningful gradient trends associated with CO_2 emissions affecting both microbenthic communities and macroalgal community structures [33, 34]. The same area was subject also to geochemical investigations conducted in order to analyze the geochemical fluxes and acoustic methodology development for their long-term monitoring [20, 35]. Similarly, innovative approaches were applied recently in the deep Pacific Ocean to investigate the fluxes in the Vailulu'u seamount caldera, using specific CO_2 probes to take samples off for further investigation onboard and on land (**Figure 18**).

Figure 17.
Examples of scientific surveys and samples collected around the coastal hydrothermal emissions discovered off Panarea Island (Eolian Archipelago, Sicily).

Figure 18.
Innovative CO_2 sampling approaches developed in the Vailulu'u crater during the NA165 cruise of E/V Nautilus conducted in deep Pacific waters on September 2024.

Further investigations revealed also the occurrence of alien species, including tropical seagrass such as *Halophila stipulacea* detected also in other areas of the Western Mediterranean Sea, as a consequence of warming conditions and seawater temperature rise detected in different coastal areas over the years [25].

4.3 Time machinery in fossilized carbon skeletons

While exploring an unnamed seamount off American Samoa during the expedition NA165 of E/V Nautilus we came through a deep coral forest where beside living organisms we observed some fossilized branches coated in ferromanganese structures. This observation leads the science team to assume these fragments are there from millennia and they went through a fossilization process over the centuries. Also, fossilized structure can act as proper time machines, revealing interesting information on the past climate history since using radiocarbon techniques is possible to predict the CO_2 level at a certain moment reconstructing the climate variability and carbonization trends (**Figure 19**).

4.4 Climate vulnerability assessment

Climate vulnerability assessments identify which habitats, species or communities may be most vulnerable based on their exposure to projected changes in the environment and their sensitivity degree to these changes. These assessments can be useful to identify areas worthy of additional research and concrete actions to reduce risks.

This approach uses information on species' life histories and distributions, projected future climate and ocean conditions to estimate vulnerability to abrupt or so significative environmental conditions. Similar approaches are used for community vulnerability assessments and require periodical observations at the same sites. This assessment can target either protected species as sea turtles or mammals, key habitats, fishes' stocks or social vulnerability of coastal and fishery communities. An example is the unprecedent mortality phenomenon in the Orbetello lagoon and in the Maremma

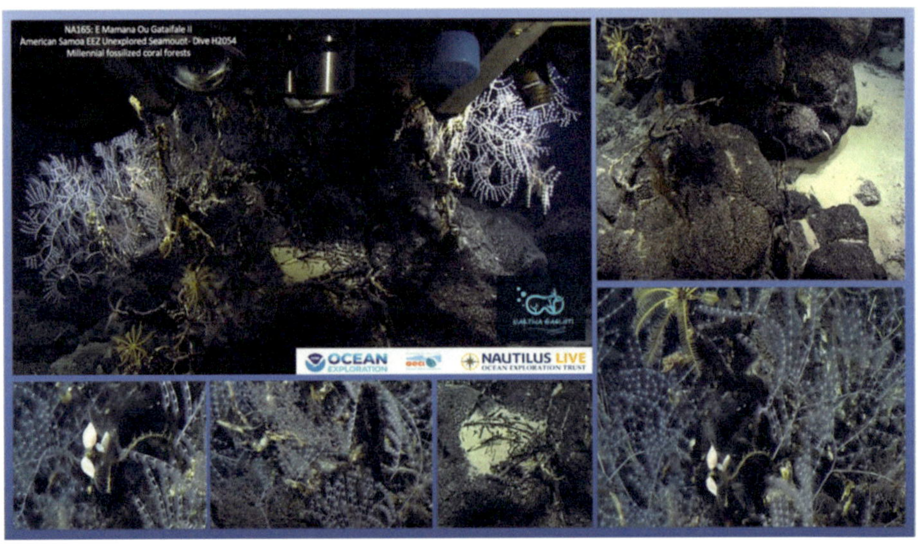

Figure 19.
Fossilized coral skeletons in American Samoa EEZ beside living coral colonies from Chrysogorgia *spp.*

Figure 20.
Dissolved oxygen and temperature daily trends measured in three stations within the Orbetello lagoon in summer 2024 (Source: https://www.arpat.toscana.it/datiemappe/bollettini/bollettino-mensile-sulla-laguna-di-orbetello-gr).

coastal area, occurred in summer 2024 along the Tyrrhenian Coast of Italy, where due to the breaking records rising temperature oxygen depletion and mass mortality events were detected with an overall depletion of 200 tons in terms of halieutic resources (**Figure 20**). It caused an estimated economical damage of 10–12 million euros. In this place, renowned as an aquaculture place where sparids are farmed for human consumption, many specimens perished to the prolonged anoxic conditions due to heat-driven oxygen depletion in their surrounding waters. During the seasonal emergency, a monitoring protocol aimed at measuring the sulfidic acid (H_2S) concentration was established, due to the concrete risk of high concentration due to the higher by production by sulfate-reducing bacteria, which, under anaerobic conditions, uses sulfates instead of oxygen in the process of mineralizing organic matter.

5. Climate resilience and adaptation strategies

5.1 The long journey toward nature restoration (alien species, coastal erosion and regime shifts affecting key ecosystems)

Many of the EU Member States still being unprepared in proper restoration plan implementation. After a long negotiation journey, in August 2024 the Nature Restoration Law (NRL) entered into force. There are so many restoration projects ongoing and EU level, but the extensive knowledge needed for large-scale landscape restoration projects even in the long term still lacking in some aspects [36]. According to the outcomes of the most recent SER Conference held in August in Tartu (Estonia), for a successful journey toward the effective implementation of the NRL goals is essential to complementary invest on scientific and experiential knowledge (meaning: stories, opinions, examples shared in common language), recognizing at the same times the key role of case studies in increasing the overall desired resilience.

The overall resilience achievement is part of the One Health approach, defined as a new model that seeks a right balance between human health and all the other biosphere components. This approach includes many different sectors and disciplines in the attempt to enhance well-being and tackle global health and environmental threats [37].

6. Upcoming initiatives and youth engagement

6.1 Global-ONCE program

The Global-ONCE program encourages a dedicated commitment to implementing ocean-based solutions that support a sustainable and resilient future. This proclamation serves as a manifesto, emphasizing the urgent need to focus on Ocean Negative Carbon Emissions (ONCE) as a critical component of global efforts to battle climate change. Global-ONCE's commitment to leveraging the underutilized capacity of our ocean to address climate threats extends beyond scientific research. It represents a dedicated pledge-to-action to protect blue carbon ecosystems and implement ocean-based solutions that promote sustainability and intergenerational equity. By integrating science, innovation and community engagement, we can proactively contribute to environmental stewardship and work toward a more sustainable and resilient future. ECOPs—including scientists, policymakers, stakeholders, non-governmental organization staff, school teachers, young ocean enthusiasts and community leaders—from all regions and countries must take action at every possible level to fully protect our ocean.

The ECOPs Declaration on Ocean Negative Carbon Emissions (ONCE) underscores the vital role of early careers and young people in addressing the pressing challenges facing our Ocean and Planet. Through advocacy, innovation and collaboration, ECOPs can drive transformative changes and secure a sustainable future for the next generations. By recognizing the importance of protecting our ocean and advocating for behavioral changes, we can overcome the challenges posed by anthropogenic climate change and pave the way for a brighter, more sustainable future. ECOPs are energetic and diverse, representing various backgrounds, cultures and experiences. By involving ECOPs from different regions and communities, we ensure that diverse perspectives are considered, leading to more inclusive and effective practical solutions (**Figure 21**).

Figure 21.
Ocean literacy world conference and launch of the Venice declaration for ocean literacy in action (June 2024).

6.2 The climate issues in the Venice declaration for ocean literacy in action

The Venice Declaration is a living document signed on the occasion of the Ocean Literacy World Conference, held in Venice (Italy) on June 2024. This event represented the first United Nations conference dedicated to ocean education, organized by the Prada Group and the UNESCO Intergovernmental Oceanographic Commission. In this occasion, 131 delegates from many different Member States affiliated to UNESCO and ocean literacy experts from around the Globe are willing to contribute to the draft designing of the "Venice Declaration of Ocean Literacy." The document is a 10-point collective document aimed at rebalancing the relationship between ocean and human beings. Among the key actions outlined there is an article specifically addressed to the climate issues. It indicates as a priority: "Raising awareness of the complex and interconnected link between ocean health, biodiversity, human well-being and the climate crisis. We must address the impact of human behaviors in the short and long term as a regenerative and destructive force. We call for a systemic approach to restore, repair, regenerate and reconnect our relationship with the ocean in all our activities—sports, recreation, work, education and government."

7. Regulatory frameworks and policies for climate action

7.1 Climate-relevant conservation measures

The United Nations Framework Convention on Climate Change (UNFCCC) is one of three conventions that was adopted at the Rio Earth Summit in 1992 to promote a sustainable planet for future generations, and was adopted in 1994 from 197 Parties. Its sister Rio Conventions are the UN Convention on Biological Diversity and the Convention to Combat Desertification. One of the aims of UNFCCC is to promote the international cooperation to fight against the challenges related to the climate crisis while ensuring sustainable economic development. The UNFCCC is the parent treaty of the Paris Agreement and the Kyoto Protocol. The ultimate objective of these agreements is to stabilize greenhouse gas concentrations in the atmosphere at a level that will prevent dangerous human interference with the climate system, in a time frame allowing ecosystems to adapt naturally fostering a sustainable development at the same time [38].

7.2 Advancing science-based policy

Recently was published the outcome of the participative process around the "Global Online Stakeholder Consultation: Inputs to Ocean Action Panels Concept Papers."

What emerged from the consultation is the transformative action that needs to happen is the Creation of Integrated Ocean-Climate Biodiversity Action Plans. Integrated action plans ensure that efforts to mitigate climate change, protect marine ecosystems, and conserve biodiversity are mutually reinforcing, leading to more effective and sustainable outcomes.

7.3 International taskforces for climate crisis and conservation-related issues

The IUCN Climate Crisis Commission's works with the aim to keep the global warming limited to 1.5 degrees above pre-industrial levels. One of the main goals is to

Figure 22.
Small-scale fishery and bycatch examples along the Med Sea Tyrrhenian coasts (Image sources: SeaPaCS Project; Martina Gaglioti).

allow society to adapt and manage climate risks in ways that enhance socio-ecological resilience, are nature positive and promote outcomes for all. The mission of CCC will be to drive the change in order to promote appropriate solutions to the climate crisis relying on the best available scientific evidence from all reputable international bodies and from indigenous knowledge systems. To achieve the mast relevant goals in these terms is important to enable a "whole of society approach" wherever applicable. Beside these more general commissions, there are other specific taskforces aimed at certain critical matters resolutions such as the IUCN resolution 107 for fishing impacts reductions on marine biodiversity, the Communication and Education Commission (CEC) devoted to the outreach and dissemination tasks and the IUCN Species Survival Commission (SSC) Coral Specialist Group having a wide conservation focus. Instead, the IUCN Species Survival Commission (SSC) works independently and with the different IUCN Units to build knowledge on the status of species and threats they face. The knowledge gathered from the people voluntary committed to these commissions is useful to provide advice, develop policies and guidelines and facilitate conservation planning development. Through this work, SSC catalyzes conservation action and enables IUCN to influence policy and assist societies in biodiversity conservation, for instance, in case of unsustainable fishing practices or damages consequent to unselective fishing method adoption (**Figure 22**) [39].

DOI: http://dx.doi.org/10.5772/intechopen.1008533

Author details

Martina Gaglioti
NOAA Scientist Ashore-Ocean Exploration Trust, Essex Village, Connecticut, USA

*Address all correspondence to: mart.gaglioti@gmail.com

References

[1] Dumitru A, Wendling L. Evaluating the Impact of Nature-Based Solutions: A Handbook for Practitioners. Publications Office of the European Union; 2021. Available from: https://data.europa.eu/doi/10.2777/11361

[2] European Commission: Directorate-General for Research and Innovation. Ocean-Climate Nexus – Understanding the Changes, Responding to the Challenges. Publications Office of the European Union; 2022. Available from: https://data.europa.eu/doi/10.2777/383194

[3] Rossi S, Bramanti L, editors. Perspectives on the Marine Animal Forests of the World. Cham, Switzerland: Springer; 2020

[4] Amadi E, Mosnier F. The Climate Myth of Deep Sea Mining. Planet Tracker; 2023. Available from: https://planet-tracker.org/wp-content/uploads/2023/06/Deep-Sea-Mining.pdf

[5] Köchy M, Hiederer R, Freibauer A. Global distribution of soil organic carbon–part 1: Masses and frequency distributions of SOC stocks for the tropics, permafrost regions, wetlands, and the world. The Soil. 2015;**1**(1):351-365

[6] Atwood TB, Witt A, Mayorga J, Hammill E, Sala E. Global patterns in marine sediment carbon stocks. Frontiers in Marine Science. 2020;7:165

[7] Baco AR, Ross R, Althaus F, Amon D, Bridges AE, Brix S, et al. Towards a scientific community consensus on designating vulnerable marine ecosystems from imagery. PeerJ. 2023;**11**:e16024

[8] von Schuckmann K, Moreira L, Grégoire M, Marcos M, Staneva J, Brasseur P, et al. 8th Edition of the Copernicus Ocean State Report (OSR8). Copernicus Publications, State Planet, 4-osr8; 2024. DOI: 10.5194/sp-4-osr8

[9] Galloway SB, Bruckner AW, Woodley CM, editors. Coral health and disease in the Pacific: Vision for action. In: NOAA Technical Memorandum NOS NCCOS 97 and CRCP 7. Silver Spring, MD: National Oceanic and Atmospheric Administration; 2009. 314 p

[10] Carpenter KE, Abrar M, Aeby G, Aronson RB, Banks S, Bruckner A, et al. One-third of reef-building corals face elevated extinction risk from climate change and local impacts. Science. 2008;**321**(5888):560-563

[11] Burke L, Reytar K, Spalding M, Perry A. Reefs at Risk Revisited. Washington, DC: World Resources Institute (WRI); 2011

[12] Lettrich MD, Asaro MJ, Borggaard DL, Dick DM, Griffis RB, Litz JA, et al. A Method for Assessing the Vulnerability of Marine Mammals to a Changing Climate. National Oceanic and Atmospheric Administration; 2019

[13] Zelli E, Ellis J, Pilditch C, Rowden AA, Anderson OF, Geange S, et al. Identifying climate refugia for vulnerable marine ecosystem indicator taxa under future climate change scenarios. Journal of Environmental Management. 2025;**373**:122635

[14] ICES. Workshop on status of vulnerable marine ecosystems in the NEAFC regulatory area for 5-year review (WKVMESTAT). ICES Scientific Reports. 2024;**6**(78):109. DOI: 10.17895/ices.pub.27109789

[15] Burgos S, Wolf J, Kaugure L, Bernoux M, Drieux E, Vasileiou I, et al. Advancing Climate Action on Agriculture and Food Security – Progress on Agriculture-Related Climate Change Negotiations and the Sharm el-Sheikh Joint Work (SJWA) from COP27 Onwards, Technical Note. Rome: FAO; 2024

[16] UNESCO-IOC. Ambition, Action, Impact: The Ocean Decade Pathway to 2030. Consolidated Outcomes of the Vision 2030 Process. Paris: UNESCO; 2024. (The Ocean Decade Series, 50)

[17] FAO, UNEP. Rethinking our Food Systems: A Guide for Multi-Stakeholder Collaboration. Food & Agriculture Org; 2023

[18] Gerber PJ, Steinfeld H, Henderson B, Mottet A, Opio C, Dijkman J, et al. Tackling Climate Change through Livestock – A Global Assessment of Emissions and Mitigation Opportunities. Rome: Food and Agriculture Organization of the United Nations (FAO); 2013

[19] Woulds C, Bell JB, Glover AG, Bouillon S, Brown LS. Benthic carbon fixation and cycling in diffuse hydrothermal and background sediments in the Bransfield Strait, Antarctica. Biogeosciences. 2020;**17**(1):1-12

[20] Di Bella L, Conte AM, Conti A, Esposito V, Gaglioti M, Ingrassia M, et al. Potential resilience to ocean acidification of benthic foraminifers living in *Posidonia oceanica* meadows: The case of the shallow venting site of Panarea. Geosciences. 2022;**12**(5):184

[21] Auriemma R, De Vittor C, Esposito V, Gaglioti M, Gambi MC. Motile Fauna associated to Cystoseira brachycarpa along a gradient of ocean acidification at a vent system off Panarea (Aeolian Islands, Italy). Biologia Marina Mediterranea. 2019;**26**:216-219

[22] Gaglioti M, Auriemma R, De Vittor C, Esposito V, Teixido N, Gambi MC. A pilot study on *Posidonia oceanica* features of a hydrothermal system off Panarea (Aeolian Islands, Italy). In: Proceedings of the 50∘ Congresso Della Società Italiana di Biologia Marina, Livorno, Italy. 2019. pp. 10-14

[23] Roche B, White PR, Bull JM, Leighton TG, Li J, Christie C, et al. Methods of acoustic gas flux inversion—Investigation into the initial amplitude of bubble excitation. The Journal of the Acoustical Society of America. 2022;**152**(2):799-806

[24] Gaglioti M, Bull JM, Roche B. Strategies for environmental monitoring of marine carbon capture and storage. In: Measuring Bubbles in the Ocean Using the Panarea CO_2 Seeps as a Natural Laboratory. STEMM-CCS Newsletter; 2020, 2018. DOI: 10.13140/RG.2.2.20420.04482

[25] Di Genio S, Gaglioti M, Meneghesso C, Barbieri F, Cerrano C, Gambi MC. Phenology and ecology of the alien seagrass *Halophila stipulacea* in its northern range limit in the Mediterranean sea. Aquatic Botany. 2021;**168**:103304

[26] World Health Organization, UNEP United Nations Environment Programme, & World Organisation for Animal Health. One Health Joint Plan of Action (2022-2026): Working Together for the Health of Humans, Animals, Plants and the Environment. World Health Organization; 2022

[27] Novaglio C. Climate change risks to marine ecosystems and fisheries. In: Projections to 2100 from the Fisheries and Marine Ecosystem Model Intercomparison Project. Vol. 707. FAO Fisheries and Aquaculture Technical Paper; 2024

[28] Jarvis C. Threats to our Ocean Heritage: Bottom Trawling. Springer Nature; 2024. p. 109

[29] Protocol K. United Nations framework convention on climate change. Kyoto Protocol, Kyoto. 1997;**19**(8):1-21

[30] Burger J. Environmental management: Integrating ecological evaluation, remediation, restoration, natural resource damage assessment and long-term stewardship on contaminated lands. Science of the Total Environment. 2008;**400**(1-3):6-19

[31] Carmona-Moreno C, Crestaz E, Cimmarrusti Y, Farinosi F, Biedler M, Amani A, et al. Implementing the Water–Energy–Food–Ecosystems Nexus and Achieving the Sustainable Development Goals. IWA Publishing; 2021

[32] Misiejuk K, Khalil M. The co-design process of an instructor dashboard for remote labs in higher education. In: Zaphiris P, Ioannou A, editors. DOI.org (Crossref). Switzerland: Springer Nature; 2024. DOI: 10.1007/978-3-031-61672-3_5

[33] Esposito V, Auriemma R, De Vittor C, Relitti F, Urbini L, Kralj M, et al. Structural and functional analyses of motile fauna associated with Cystoseira brachycarpa along a gradient of ocean acidification in a CO_2-vent system off Panarea (Aeolian Islands, Italy). Journal of Marine Science and Engineering. 2022;**10**(4):451

[34] Berlino M, Mangano MC, Di Bona G, Lucchese M, Terzo SMC, De Vittor C, et al. Functional diversity and metabolic response in benthic communities along an ocean acidification gradient. Marine Environmental Research. 2024;**198**:106520

[35] Louvari MA, Auriemma R, Bigi S, Caruso C, Conti A, De Vittor C, Di Bella L. Preliminary data on recent benthic foraminiferal assemblages from the shallow-water hydrothermal system of the Aeolian Arc (Tyrrhenian Sea).

[36] Hering D, Schürings C, Wenskus F, Blackstock K, Borja A, Birk S, et al. Securing success for the nature restoration law. Science. 2023;**382**(6676):1248-1250

[37] Mackenzie JS, Jeggo M. The one health approach—Why is it so important? Tropical Medicine and Infectious Disease. 2019;**4**(2):88

[38] Kuyper J, Schroeder H, Linnér BO. The evolution of the UNFCCC. Annual Review of Environment and Resources. 2018;**43**(1):343-368

[39] IUCN. Home page [Internet]. IUCN. Available from: https://iucn.org/our-union/commissions/iucn-speciessurvival-commission-2021-2025

Chapter 4

Implementing Carbon Storage Technology to Combat Climate Change

Ganesh Thakur, Xuejia Du and Sabyasachi Prakash

Abstract

Injecting CO_2 into depleted oil reservoirs is a key strategy for mitigating excess atmospheric carbon. In India, where Carbon Capture, Utilization, and Storage (CCUS) incentives are limited, CO_2 EOR (Enhanced Oil Recovery) serves as a commercially viable method for carbon storage. Effective reservoir management (RM) is critical to ensuring the success of CO_2 injection, as it integrates insights from primary, secondary, and EOR/EOR+ phases. This study introduces an integrated carbon storage development strategy tailored for a mature Indian oilfield, employing both analytical and numerical tools to conduct a thorough analysis of a depleted stacked pay reservoir. Using over 30 years of dynamic field data, we identified a storage potential exceeding five million metric tons, alongside an incremental oil recovery factor of 11% of the original oil-in-place (OOIP). Additionally, eliminating the waterflooding stage enhances storage capacity by an estimated 0.5 million tons, while ongoing aquifer water production could contribute approximately 0.35 million tons of storage annually. This approach highlights the significant carbon storage potential in mature fields, emphasizing the importance of integrated RM, particularly in regions lacking robust CCUS policies.

Keywords: reservoir management, CCUS, carbon storage, EOR, reservoir simulation, reservoir engineering

1. Introduction

1.1 Overview of worldwide energy production and consumption

Energy production and consumption are fundamental elements of modern economic growth. The economic growth has a strong positive impact on energy consumption both in long-run and short-run [1, 2]. Fossil fuels such as oil, natural gas, and coal still prevail as the main energy source of power production with a share of 79.5% to the global energy demands as of 2018 [3]. While energy sources are diversifying and renewable energy emerge as alternatives, liquid fossil fuels remain a vital energy source, specially in developing countries. The U.S. Energy Information Administration's latest International Energy Outlook 2017 (IEO2017) projects that world energy consumption is expected to increase by 28% between 2015 and 2040 [4], as shown in **Figure 1**. The majority of this increase is anticipated to come from countries where demand is driven by strong economic growth, especially in Asia (such as China and India), which will account for more than 60% of the global increase in energy consumption from 2015 through 2040.

Despite the fact that IEO2017 projects the nonfossil fuels (renewables and nuclear) would grow more quickly than fossil fuels, fossil fuels still account for more than three-quarters of world energy consumption through 2040 [4]. The intense energy demand from fossil fuels will inevitably lead to increased CO_2 emissions. Annual fossil CO_2 emissions continue to rise among the world's major GHG (green house gas) emitters. As highlighted by Bloomberg news, "Climate change is not a problem with a single solution. And it's not a challenge that any one group – governments, companies, scientists or individual citizens – can solve alone." Collaborative efforts are essential to building a healthier and more sustainable future for generations to come. By applying diverse technologies, including solar, wind, geothermal, nuclear, extended batteries, and hydrogen, along with robust government support, dedicated companies, universities, research centers, and regulatory agencies, we have a significant opportunity to address this global challenge.

1.2 Rising CO_2 emissions and the need for mitigation strategies

When it comes to the amount of GHG that is present in the atmosphere, CO_2 is the dominant contributor, contributing to 60% of global warming effects. CO_2 emissions rose from 1.95 billion metric tonnes in 1900 to 36.3 billion metric tonnes in 2021 as a result of the modern society's rapid development [1]. Each year, more than 30 billion tonnes of CO_2 is further emitted into the atmosphere, aggravating the climate change issue [5]. The average atmospheric concentration of CO_2 has increased from 280 ppm to 370 ppm since the start of industrial revolution in around 1850. Consequently, during the same time period, the average global temperature has increased by 0.6 to 1°C [6]. According to the International Panel on Climate Change (IPCC), the atmosphere might contain up to 570 ppm CO_2 by the year 2100, which would result in an increase in the average global temperature of around 1.9°C and a 38 m rise in sea level [6]. Continued uncontrolled greenhouse gas emissions may lead to sea level increase and species extinction. Therefore, reducing CO_2 levels in the atmosphere is crucial to

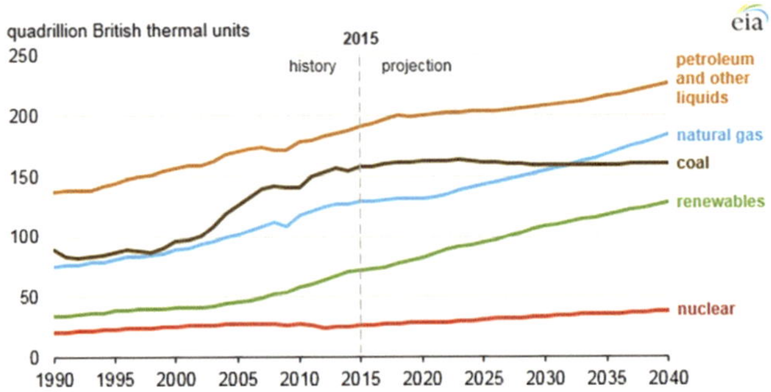

Figure 1.
World energy consumption by energy source (1990–2040) [4].

mitigating climate change, since excessive CO_2 emissions contribute to global warming and the resulting catastrophic environmental and economic impacts.

1.3 Definition and significance of CCUS

In 1970s, the idea of capturing CO_2 emissions to mitigate climate change began to take shape. Early research focused on understanding the feasibility of capturing CO_2 from industrial processes and storing it underground. The U.S. Department of Energy (DOE) started funding research into CO_2 capture in 1977, laying the groundwork for future CCS technologies. Cesare Marchetti, an Italian physicist, proposed a method for controlling CO_2 levels in the air, in which the CO_2 is collected at appropriate concentrated emission points and transferred to the deep sea or underground caverns for permanent storage [7]. During the 1980s, the concept of CCS gained more attention as concerns about global warming and greenhouse gas emissions grew. Research institutions and scientists began exploring various methods for capturing and storing CO_2. Norwegian scientist Erik Lindeberg of the Norwegian Petroleum Directorate (NPD) and the SINTEF research institute were among the early advocates for CO_2 storage, particularly in deep saline aquifers, and depleted oil and gas fields. The International Energy Agency (IEA) and other organizations began to promote CCS as a critical technology for reducing CO_2 emissions. The establishment of the Global CCS Institute in 2009 marked a significant step in promoting and advancing CCS technologies worldwide. The IPCC recognized CCS as an important mitigation option in its Third Assessment Report (2001) and subsequent reports. The purpose of the IPCC report was to alert the policy-makers, engineers, and researchers about global warming due to CO_2 emissions and the need to develop practical solutions to deal with this problem. The concept of Carbon Capture, Utilization, and Storage (CCUS) emerged as a way to enhance the economic viability of CCS by finding ways to use captured CO_2. Since then, CCUS has been widely recognized as an effective technology for reducing atmospheric CO_2 levels and is increasingly being used in industrial carbon capture.

1.4 Global perspective on CCUS projects and their growth potential

CCUS involves several methods across its three main stages: capture, storage, and utilization, each with its own set of technologies and applications (**Figure 2**). The current technologies of CO_2 capture includes DAC (direct air capture), pre-combustion technology, post-combustion technology, oxy-fuel combustion technology, and chemical looping technology. The current global CO_2 captured capacity is only around 0.1% (~40 million metric tonnes per year) of total global emissions (~36 billion tonnes per year). This is due to two main reasons: first, the high cost for extensive infrastructure (pipelines, storage sites), and regulatory hurdles are significant challenges to scaling up CO_2 capture capacity; second, the technology to accelerate the deployment of CCUS technologies and integrate them into more industries is challenging to increasing global capacity. The CO_2 capture alone accounts for more than 70% of all operating expenses of CCS [8].

CO_2 utilization refers to processes that convert captured CO_2 into valuable products, thereby reducing the amount of CO_2 released into the atmosphere. Several methods are used for CO_2 utilization, each with varying levels of maturity, economic viability, and market potential. Common methods include Enhanced Oil Recovery (EOR), chemical conversion, biological conversion, and industrial

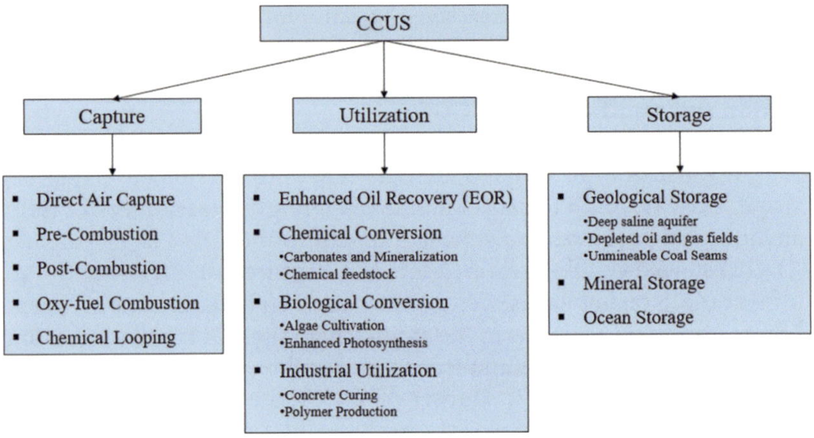

Figure 2.
CCUS categories.

utilization. Among all these methods, EOR is currently the most widely used method of CO_2 utilization, accounting for the majority of CO_2 used globally. It's a mature technology with decades of application in the oil and gas industry. EOR operations can utilize large quantities of CO_2, often in the range of millions of tons per project annually. Approximately 30–40 million metric tons of CO_2 are used annually for EOR globally [9]. The potential for CO_2 use in EOR is significantly higher, especially with the expansion of CO_2 capture and transport infrastructure. Some estimates suggest that if fully utilized, EOR could consume up to 140–370 million metric tons of CO_2 annually by 2050, depending on oil prices and technological advances.

CO_2 storage is trying to securely store CO_2 to prevent it from entering the atmosphere. The common methods of CO_2 storage include geological storage, mineral storage, and ocean storage. Geological storage is the most mature and proven method, with numerous projects worldwide demonstrating its effectiveness and lowest leaking risk for thousands of years. The global capacity for geological CO_2 storage is substantial, but the amount of CO_2 currently being stored annually is still relatively modest compared to the potential capacity, approximately 20–25 million metric tons of CO_2 are being stored annually worldwide [8, 9].

The first commercial CCS project, Sleipner, started in 1996 in the North Sea, operated by Statoil (now Equinor). The project involved capturing CO_2 from natural gas processing and storing it in a saline aquifer. There are 65 large-scale CCS/CCUS demonstration projects worldwide at the moment. Of these, 26 are operational, 2 are temporarily suspended, 3 are under construction, 13 are in the front-end engineering design, and 21 are in the early research stage. These projects are primarily concentrated in the United States, China, Canada, Norway, Australia, and the United Arab Emirates [10]. **Table 1** lists several outstanding projects. These projects vary in size, scope, and application, demonstrating the diverse potential of CCUS technologies, and there is a tremendous growth anticipated in this area.

As fossil fuels continue to dominate the global energy mix and stringent greenhouse gas regulations become more widespread, CCUS has emerged as a crucial technology for achieving low-carbon energy solutions. For nations with large energy demands, such as India and China, CCUS offers significant potential to enhance environmental sustainability while bolstering energy self-reliance. Driven by financial incentives for carbon capture, many operators are increasingly exploring CO_2 EOR and storage projects. At the University of Houston (UH), our research focuses on CO_2 EOR in both conventional and unconventional reservoirs, incorporating laboratory studies, reservoir characterization, dynamic simulation, and integrated reservoir-well facilities analysis. UH's fundamental research is further strengthened by field studies, particularly in collaboration with Oil India, addressing industrial engineering challenges in CO_2 EOR/Storage ventures.

No.	Name	Type	Location	Details
1	Petra Nova [11]	Capture & Utilization (EOR)	Texas, USA	Captures CO_2 from a coal-fired power plant and transports it to an oil field for EOR, with a capacity to capture up to 1.4 million tonnes of CO_2 annually.
2	Boundary Dam [11]	Capture & Storage	Saskatchewan, Canada	The first large-scale CCS project on a coal-fired power plant, capturing around 1 million tonnes of CO_2 annually. The captured CO_2 is used for EOR and stored in a deep saline aquifer.
3	Illinois Industrial Carbon Capture and Storage [12]	Capture & Storage	Illinois, USA	Captures CO_2 from an ethanol production facility and stores it in a sandstone formation. The project can store up to 1 million tonnes of CO_2 annually.
4	Sleipner [13]	Capture & Storage	North Sea, Norway	Operational since 1996, Sleipner captures CO_2 from natural gas processing and stores it in a saline aquifer beneath the North Sea. It stores about 1 million tonnes of CO_2 annually.
5	Northern Lights [13]	Capture & Storage	North Sea, Norway	Part of the Norwegian full-scale CCS project, aims to capture and store CO_2 from various industrial sources. The first phase will store up to 1.5 million tonnes of CO_2 annually.
6	Gorgon CO_2 Injection Project [14]	Capture & Storage	Western Australia	Captures CO_2 from a natural gas processing plant and injects it into a deep saline aquifer. The project aims to store up to 4 million tonnes of CO_2 annually.
7	Tomakomai CCS Demonstration Project [15]	Capture & Storage	Hokkaido, Japan	Captures CO_2 from a hydrogen production facility and stores it in offshore geological formations. The project aims to store up to 100,000 tonnes of CO_2 annually.
8	Al Reyadah [16]	Capture & Utilization (EOR)	Abu Dhabi	Captures CO_2 from a steel plant and uses it for EOR in nearby oil fields. The project can capture up to 800,000 tonnes of CO_2 annually.
9	Sinopec's Qilu Petrochemical CCUS Project [10]	Capture & Utilization (EOR)	China	This is one of the largest CCUS projects in China. The project is designed to capture and store approximately 1 million metric tons of CO_2 annually.

Table 1.
Examples of CCUS projects around world.

2. Enhanced oil recovery (EOR) with CO_2 injection

2.1 History and development of CO_2 EOR

Enhanced oil recovery (EOR) refers to a procedure in which a substance, such as CO_2, is injected into a reservoir to repressurize rock formation and to release any trapped oil/gas. During the CO_2 EOR process, the injected CO_2 can reduce oil viscosity, increase oil mobility, allowing oil to flow more easily from the otherwise hard-to-recover formation. The oil-CO_2 mixture is then pumped to the surface, where CO_2 is separated from the oil and reinjected into the reservoir to repeat the process (**Figure 3**). This process often yields more barrels per reservoir compared to traditional oil-recovery methods. CO_2-EOR is a well-established practice that enables the recovery of additional oil beyond secondary and primary recovery techniques.

The first commercial CO_2 injection project was the SACROC (Scurry Area Canyon Reef Operators Committee) flood in 1972 in the Permian Basin of US. The success of the SACROC project demonstrated the feasibility of CO_2 EOR and spurred interest in the technology. Since then, multiple commercial-scale CO_2 floods have been performed in the Permian Basin. By the late 1970s and early 1980s, several other oil fields in the region adopted CO_2 injection, benefiting from the proximity to natural CO_2 sources like the McElmo Dome and Sheep Mountain fields, which provided a reliable supply of CO_2. Advances in reservoir management, including better understanding of CO_2 behavior in the reservoir, improved the efficiency of CO_2 EOR. Techniques such as alternating water and CO_2 injection (known as WAG – Water Alternating Gas) were developed to optimize oil recovery and CO_2 utilization. While the United States remained the leader in CO_2 EOR, the technology began to be adopted in other oil-producing regions, including Canada, the Middle East, and China. Recent years have seen further advances in monitoring and verification technologies, ensuring that CO_2 injected during EOR remains securely stored in the reservoir. Techniques such as 4D seismic imaging, tracer studies, and reservoir simulation models have improved the ability to monitor CO_2 movement and storage

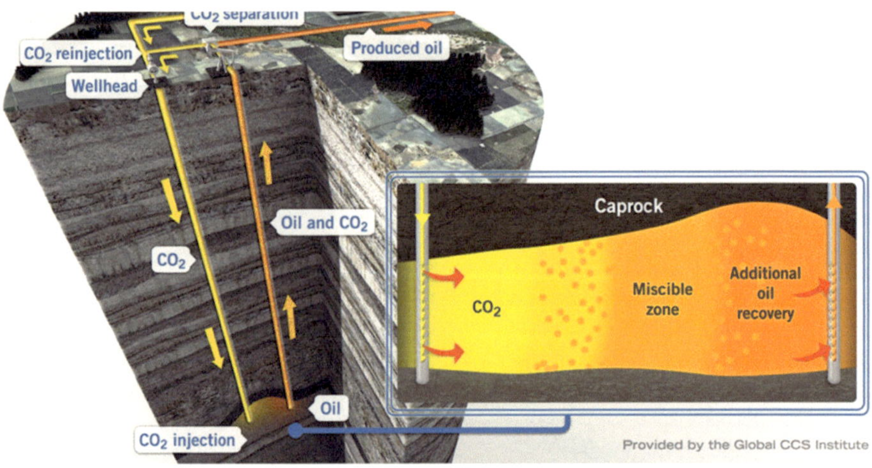

Figure 3.
CO_2 injection process [17].

DOI: http://dx.doi.org/10.5772/intechopen.1006915

efficiency. As of now, CO_2 EOR is one of the popular techniques to enhancing oil recovery, occupying around 20% of 1120 worldwide EOR projects [18].

2.2 Dual benefits of CO_2 injection: enhanced oil recovery and carbon storage

CO_2 EOR has evolved to a mature technology with significant implications for both energy production and carbon management. The utilization of CO_2 in EOR can significantly improve oil recovery, with the potential to recover 15–25% of the OOIP in light or medium oil fields nearing depletion due to flooding [19]. If reinjection is not applied, approximately 60% of injected CO_2 can be trapped in the reservoir at the CO_2 breakthrough [20]. Recently, there have been efforts to incorporate geological storage as another objective [21, 22]. A notable example is the Weyburn oilfield (Canada), the first project to combine CO_2 EOR with storage, which boasts a storage capacity of approximately 25 million tonnes [21]. Numerous studies have since explored the co-optimizing of CO_2 EOR and storage [23, 24]. However, many of these studies rely on simplifying assumptions and often overlook practical challenges in brownfields, such as uneven reservoir pressure, varying fluid saturations, and high free gas saturation. Overcoming these challenges is critical for the successful implementation of CO_2 storage process.

50 years history of CO_2 EOR reflects the broader trends in the energy industry, from the initial focus on maximizing oil production to the current emphasis on sustainability and carbon sequestration. As technology continues to advance and the global focus on reducing carbon emissions intensifies, CO_2 EOR is likely to play an increasingly important role in both the oil industry and climate change mitigation efforts.

2.3 Case study: SACROC project

The SACROC Unit project is one of the most significant and pioneering CO_2 EOR projects in the world. Located in the Permian Basin of West Texas, it is one of the oldest continuously operated CO_2 EOR projects and has played a crucial role in demonstrating the viability of CO_2 injection for enhanced oil recovery. The oil field is situated within the Canyon Reef formation, which is a carbonate reservoir and conducive to oil production. It has an estimated OOIP of just under 3 billion barrels. The reservoir ranges from 200 ft. gross thickness in the south up to 900 ft. in the north where the limestone matrix averages 8% porosity and 20 md permeability. SACROC Unit was discovered in 1948 and rapidly developed by 1951, with 1671 producing wells drilled. In 1972, SACROC became the site of the first large-scale commercial CO_2 EOR project. This marked a major milestone in the development of enhanced oil recovery techniques. In the early years, the amount of CO_2 injected was relatively modest as the technology and infrastructure were being developed. By the 1990s, the SACROC Unit had become one of the largest CO_2 EOR projects, with millions of tons of CO_2 injected into the reservoir. It is estimated that over 200 million tons of CO_2 have been injected into the SACROC Unit since the project began. This figure includes both naturally sourced CO_2 and, more recently, CO_2 captured from industrial sources. In recent years, the annual CO_2 injection rates at SACROC have been substantial, with estimates ranging from 2 to 4 million tons of CO_2 per year. The project has added millions of barrels of incremental oil production through CO_2 injection, with additional recovery factors in the range of 10–15% beyond conventional methods [25].

3. Reservoir management for CO_2 injection

3.1 Definition and importance of intelligent reservoir management

The concepts and processes of reservoir management were proposed and established by Satter and Thakur in 1990s [26–28] and implemented by Du and Thakur in 2024 [29]. They emphasized the importance of integrating different disciplines such as reservoir characterization, fluid and their behavior in the reservoir, drilling and operating wells, surface processing of the fluids, facilities, water system, and field operation to benefit the most of each project. Furthermore, they highlighted that reservoir management is not confined to initial engineering and geologic evaluations, economic justifications, and project approval by management. Rather, these ongoing activities span the period from before the commencement of the waterflood until the waterflood becomes uneconomical or transitions to EOR techniques.

Intelligent reservoir management (IRM) is an advanced approach to optimizing oil recovery through the integration of multidisciplinary expertise, real-time data, and cutting-edge technologies. It combines traditional reservoir management practices with modern innovations like intelligent completions, 4D seismic monitoring, and advanced data analytics to make more informed decisions that maximize oil recovery and economic value. Intelligent reservoir management across primary, secondary, and tertiary recovery phases enables a comprehensive understanding of the key physical properties and mechanisms that govern oil recovery. A well-managed reservoir is also better positioned to maximize the benefits of CO_2 injection, achieving the dual objectives of enhanced oil recovery and carbon storage.

3.2 Role of IRM in CO_2 EOR projects

IRM can be effectively applied in CO_2 injection projects, especially in CO_2 EOR operations. For example, the Weyburn-Midale CO_2 EOR project in Saskatchewan, Canada, is one of the world's largest and most successful CO_2 EOR projects. IRM techniques are extensively applied to optimize CO_2 injection and maximize oil recovery. The key IRM applications in the Weyburn-Midale project include [22, 30]:

1. Real-time monitoring and data integration:

    a. *4D seismic monitoring*: 4D seismic surveys are repeated over time to monitor the movement and distribution of CO_2 within the reservoir. It helps understand how CO_2 is sweeping through the reservoir and identifies areas where CO_2 might be bypassing oil, enabling operators to adjust injection strategies in real-time.

    b. *Intelligent wells*: intelligent well completions equipped with sensors and downhole control devices allow for real-time monitoring of reservoir conditions, such as pressure, temperature, and fluid composition. This data is transmitted to the surface, and integrated with reservoir models to optimize CO_2 injection and production rates.

2. Reservoir simulation and modeling:

    a. *Advanced reservoir simulation*: sophisticated reservoir simulation models that incorporate geological, geophysical, and production data are employed which predict the response of CO_2 injection, helping plan the optimal injection patterns and volumes.

b. *History matching*: reservoir simulation models are calibrated using historical production and injection data to improve their accuracy. This allows for better prediction of future reservoir performance and more effective management of CO_2 injection.

3. Optimization of CO_2 injection:

   a. *Water alternating gas (WAG) strategy*: IRM techniques optimize the timing and volumes of CO_2 and water injected to improve sweep efficiency and maximize oil recovery.

   b. *Adaptive CO_2 injection*: based on real-time data and model predictions, operators can adjust the CO_2 injection rates, pressures, and locations to target specific areas of the reservoir, ensuring that CO_2 is used efficiently and oil recovery is maximized.

4. Environmental and economic management:

   a. *CO_2 sequestration monitoring*: IRM monitors the long-term storage of CO_2 in the reservoir, ensuring that it remains securely trapped and does not leak into the atmosphere.

   b. *Economic optimization*: by continuously optimizing CO_2 injection and production strategies, the project maximizes oil recovery while minimizing operational costs.

The application of IRM in the Weyburn-Midale CO_2 EOR project has led to significant increases in oil recovery and has extended the life of the reservoir. The project has successfully injected and stored millions of tons of CO_2, making it a model for how IRM can be used to enhance both oil recovery and carbon sequestration, ensuring the project's long-term economic and environmental sustainability.

4. Practices, challenges, and advanced technologies in CO_2 EOR

4.1 Current state of the art of CO_2 EOR technologies

As of 2024, the status of CO_2 EOR remains both promising and complex. CO_2 EOR continues to be a critical technology in the oil industry, particularly in the United States, where it plays a significant role in boosting oil production. Over nearly 70 years, the CCUS-EOR industrial system in the United States has been mature, continuously expanded, and now remains stable (**Figure 4**) [31].

The Advanced Resources Internation has updated the survey of U.S. CO_2 EOR active projects as the end of 2022 [32]. There are 139 projects in total with 1.9 Bcf/d of CO_2 supply (**Figure 5**). They claimed that a rebound in CO_2 EOR oil production and increased CO_2 supply may happen in the following years. Furthermore, as successful implementation of 45Q tax incentives continues, CO_2 EOR remains an attractive option for generating revenue while storing CO_2 emissions.

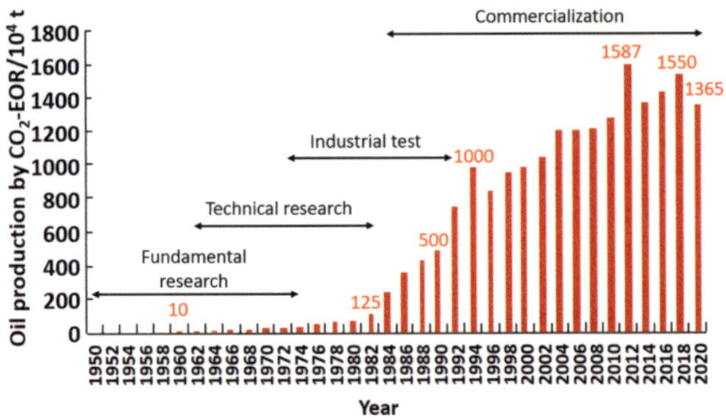

Figure 4.
Oil production by CO_2 EOR over the years in the U.S. [31].

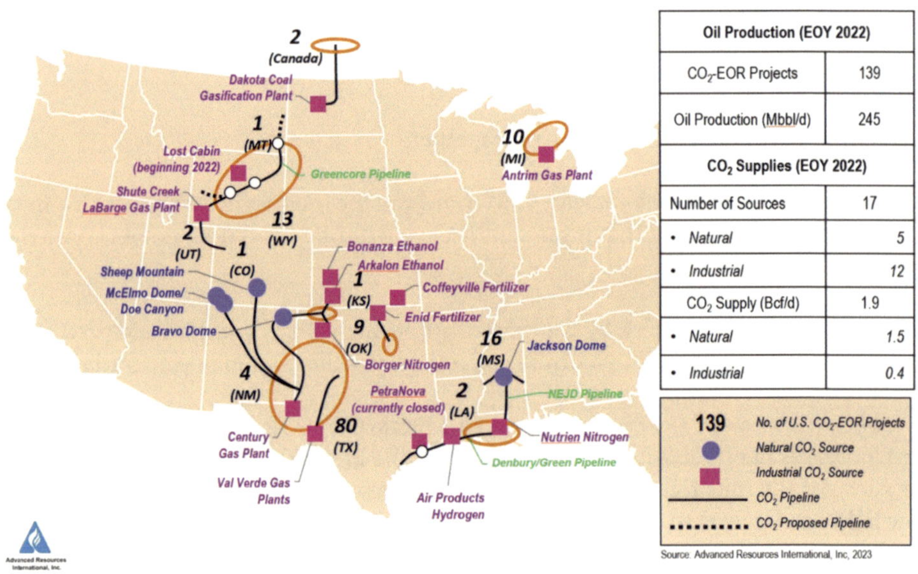

Figure 5.
U.S.CO_2 EOR active projects at 2022 [32].

A comprehensive system has been developed encompassing several key technologies: large-scale CO_2 capture, long-distance supercritical CO_2 pipeline transmission, advanced reservoir engineering for CO_2 flooding, and safety monitoring for large-scale storage. The injection, production, and surface engineering equipment are designed for simplicity, efficiency, and high automation. Ongoing advancements in dynamic monitoring allow for timely optimization and adjustments. Additionally, gas recycling technologies have been implemented to improve overall project efficiency and achieve the goal of a closed-loop, zero-emission CCUS-EOR process.

4.2 Practical challenges and uncertainties

Despite its widespread adorption in various regions and multiple benefits, CO_2 EOR still faces several challenges:

1. *CO_2 supply and cost*: a major challenge is the availability and cost of CO_2. For CO_2 EOR to be economically viable, a reliable and affordable supply of CO_2 is essential. This often requires the development of infrastructure for CO_2 capture, transportation, and storage, which can be capital-intensive.

2. *Reservoir heterogeneity*: the effectiveness of CO_2 EOR can be limited by the heterogeneity of the reservoir. Variations in rock properties, such as permeability and porosity, can lead to uneven CO_2 distribution, reducing the overall efficiency of the process.

3. *CO_2 breakthrough*: another challenge is the early breakthrough of CO_2 at production wells, which can reduce the efficiency of the oil recovery process. Managing and mitigating CO_2 breakthrough requires careful reservoir management and can increase operational complexity.

4. *Environmental and regulatory concerns*: while CO_2 EOR can contribute to carbon sequestration, there are concerns about the long-term stability of CO_2 storage and potential leakage. Additionally, regulatory frameworks need to be robust to ensure the safe and effective implementation of CO_2 EOR projects.

5. *Economic viability*: the economic viability of CO_2 EOR is highly dependent on oil prices. Fluctuating oil prices can affect the profitability of CO_2 EOR projects, making it challenging to secure long-term investments.

6. *Technical challenges*: Technical challenges related to CO_2 injection, such as controlling the injection rate, managing the mobility of CO_2 within the reservoir, and optimizing the sweep efficiency, are critical for the success of CO_2 EOR projects.

These challenges require ongoing research and development to improve the efficiency and cost-effectiveness of CO_2 EOR while ensuring environmental safety and compliance with regulations.

4.3 New technologies in predicting oil production and CO_2 storage

In recent years, data science technologies have emerged as cutting-edge methodologies for predicting oil production performance and estimate CO_2 storage capacity, while simultaneously reducing high computational time and associated costs. The advancement of computational capabilities have facilitated the development of surrogate models or proxy models in subsurface modeling [23]. Various machine learning algorithms are employed to construct these models for the prediction and optimization of hydrocarbon production in petroleum engineering. The application of proxy models can be treated as a modern approach to deliver time- and cost-effective alternatives for predicting CCUS performance and optimization [24].

These technologies are revolutionizing CO_2 EOR projects by enhancing various aspects of the process. They are used for reservoir characterization, where machine learning techniques like clustering, classification, and regression analyze large datasets to identify patterns in reservoir properties, leading to a better understanding of reservoir behavior and optimized CO_2 injection strategies. Additionally, predictive modeling with neural networks, random forests, and support vector machines forecasts outcomes such as oil recovery rates and CO_2 breakthrough times, enabling real-time decision-making [33]. Surrogate models play a crucial role in fast simulations, allowing for rapid evaluation of different CO_2 injection strategies and conducting sensitivity analyses while also quantifying uncertainties to improve risk management. These models are often integrated with machine learning to enhance predictive accuracy. Furthermore, advanced monitoring and control systems, powered by real-time data processing, enable continuous monitoring and dynamic adjustments to the EOR process, optimizing injection parameters and improving recovery efficiency. Ultimately, the combination of these technologies leads to the development of sophisticated decision support systems, providing operators with actionable insights and recommendations, which significantly enhance decision-making under complex conditions and uncertainty.

The integration of data science, machine learning, and surrogate models into CO_2 EOR projects represents a significant advancement in the field. These technologies enable more accurate predictions, better optimization, and enhanced decision-making, ultimately leading to more efficient and effective CO_2 EOR operations. As these technologies continue to evolve, their impact on CO_2 EOR is expected to grow, contributing to improved oil recovery and more effective carbon sequestration.

4.4 Future trends and innovations in CO_2 EOR technology

The potential impact of CO_2 EOR is increasingly viewed as a matter of timing rather than feasibility. The scientific principles underpinning CO_2 EOR are well-established within the field, and there is an ample supply of CO_2 emissions from various sources, as well as numerous oil reservoirs capable of accommodating these CO_2 volumes. The critical factor determining the adoption of CO_2 EOR technologies is thus not the technical feasibility but rather the economic considerations. Key economic determinants include the fluctuating price of oil, the capital investment required for infrastructure development, tax policies related to CO_2 emissions, the costs associated with capturing CO_2 from emitters, and the pace at which new technologies evolve. While forecasting these parameters can be challenging, establishing a robust decision-support platform could significantly accelerate the deployment of CO_2 EOR and sequestration technologies. Such a platform would aid in navigating the uncertainties associated with these economic factors, thereby facilitating more timely and informed decision-making processes. As technological advancements and policy frameworks continue to evolve, their impact on the economic viability of CO_2 EOR will become increasingly pivotal in shaping future trends in carbon management and oil recovery strategies.

5. Case study: CO_2 EOR and storage in Indian oilfields

5.1 Background of the project

India, as the fastest-growing major economy and the third-largest emitter of CO_2 globally, faces a critical challenge in balancing its escalating energy demands with its climate change commitments. Given this context, it is imperative for India to adopt and advance technologies that minimize carbon emissions and reduce the overall carbon footprint. The urgency of this transition is underscored by the need to reconcile economic growth with sustainable environmental practices, making the development and implementation of low-carbon technologies a priority for achieving both energy security and climate objectives. The University of Houston has collorated with one oil company in India and aim to rejuvenate production from a mature oil reservoir while simultaneously sequestering CO_2, addressing both energy and environmental concerns. The target area is the Nahorkatiya Main field, located in the central part of Upper Assam shelf foreland Basin, Northeast India. This oilfiled has produced for 60 years (primary and secondary recovery) and the reservoir has been significantly depleted. For more information of reservoir description, readers can look into the reference written by Chen et al. [34].

5.2 Methodology: reservoir screening, pilot design, and implementation phases

Detailed studies of this project include laboratory and simulation studies, reservoir characterization, conformance control and sweep efficiency, optimu pattern of wells and placement of fractures, integrated reservoir-well-facilities studies, and effective reservoir management in CO_2 EOR and CO_2 storages. This can be catergorized in four phases (**Figure 6**):

- Phase 1 – Reservoir Identification: The University of Houston has identified potential reservoirs for CO_2 EOR feasibility studies from an initial pool of 50 candidates, utilizing patented advanced reservoir screening techniques. To qualify as a suitable candidate for CO_2 EOR, a reservoir must exhibit a remaining oil saturation greater than 40%. Essential parameters for evaluation include mobility ratio, minimum miscibility pressure, and bubble point pressure values, alongside the geological characters of the reservoir, such as lateral and vertical facies variations. Additionally, the effective management of the reservoir is a crucial factor in determining its suitability for CO_2 EOR.

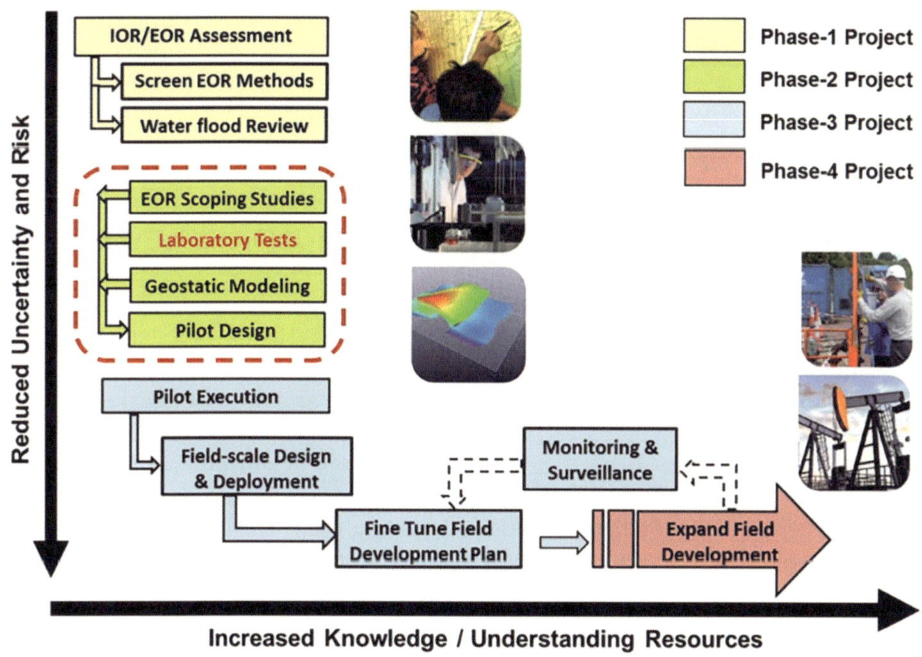

Figure 6.
The development stages of a typical EOR project [35].

- Phase 2 – CO_2 EOR Pilot Design: The pilot design phase encompasses several key components: laboratory studies, simulation studies, and pilot design. Laboratory studies involve three-dimensional geological modeling, history matching, and CO_2 EOR simulations. Three-dimensional geomodelling includes comprehensive reservoir mapping, petrophysical analysis, and static modeling. The MMP, defined as the lowest pressure at which the injected gas becomes miscible with the reservoir fluid, is determined through slim tube experiments. Additional laboratory tests, such as swelling tests, asphaltene precipitation tests, and core flooding, are conducted to evaluate the dynamics of the reservoir and assess EOR applicability. Dynamic simulation entails rigorous history matching of pressure and production data, followed by predictions of water injection and CO_2 EOR outcomes. The CO_2 storage capacity of the reservoirs is analyzed, and optimal pilot injection patterns are established, considering both economic and facility-related aspects. Collaborations with local industrial sources for anthropogenic CO_2 are also pursued.

- Phase 3 – CO_2 EOR Pilot Implementation: To date, UH-OIL collaboration has developed two CO_2-EOR pilot designs, and reservoir management operations for preparation of CO_2 EOR are ongoing. The Indian authorities have approved CO_2 injection for the first pilot project. Encouraged by the progress of this study and supported by government incentives, other operators have demonstrated significant interest and activity in similar initiatives. The successful implementation of these fieldwide projects is anticipated to contribute substantially to India's energy and environmental objectives.

- Phase 4 – Post-Pilot Evaluation and Field Development: Upon completion of the pilot test, field data, including production and pressure measurements, will be analyzed to assess the performance of the CO_2 EOR process. The reservoir model will be continuously updated and refined based on the accumulating results from the pilot test. If the pilot demonstrates both technical feasibility and economic viability, a broader, field-wide CO_2 EOR injection strategy will be implemented. This phase involves scaling up the successful techniques and methodologies from the pilot to a larger operational context, thereby extending the benefits observed at the pilot scale to the entire field.

5.3 Results and findings: incremental oil recovery and CO_2 storage potential

In this project, a geomodel was developed using various data sources, including seismic and petrophysical information, and was further refined through dynamic data analysis. The numerical simulation model was calibrated through a history-matching study, which integrated production, pressure, and petrophysical data. This calibrated model was then employed to optimize parameters for CO_2 injection. An integrated carbon storage-focused development strategy was formulated for a mature oilfield, leveraging a combination of analytical and numerical tools to perform a comprehensive analysis of a depleted stacked pay reservoir. The study utilized real field data accumulated over more than 30 years of performance. The reservoir has a storage capacity of five million metric tonnes and an incremental oil recovery factor of 11% of the original oil-in-place. The elimination of the waterflooding stage contributes an additional 0.5 million tonnes of storage capacity, while continuous aquifer water production has the potential to add approximately 0.35 million tonnes annually. Furthermore, ongoing aquifer water production also contributes to CO_2 storage on an annual basis. This research underscores the significant potential for carbon storage in over 50 fields within the region, with many of these reservoirs underlain by extensive aquifers that offer additional CO_2 storage potential.

The conclusions drawn from this project are as follows:

1. *Integrated reservoir characterization*: utilizing petrophysical and primary phase dynamic data (such as well logs, core sample data, production, and pressure data) validates the continuity of sand formation and identifies optimal injection locations for CO_2. This methodology enables reservoir managers to design effective carbon storage projects even in the absence of prior injection data.

2. *Applicability to mature oil fields*: The methodologies developed in this study can be readily adapted to other mature oil fields with similar geological characteristics. The substantial storage potential in both the oil reservoir and underlying aquifer renders carbon storage projects economically viable, particularly when coupled with increased oil production through EOR.

5.4 Impact on India's energy and environmental goals

This study has significant implications for India's energy sector and environmental goals. It demonstrates the technical feasibility of CO_2 EOR in India's specific geological settings, which differ from the well-studied reservoirs in places like the United States. The project provides a pathway to rejuvenate mature oil fields and extend their productive life, potentially boosting India's domestic oil production. At the same time, it establishes a methodology for identifying and implementing CCUS projects in India, contributing to the country's efforts to reduce carbon emissions.

6. Lessons learned and best practices

6.1 Key takeaways from successful CO_2 EOR projects

A crucial lesson from successful CO_2 EOR projects is the importance of thoroughly understanding reservoir characteristics. Detailed geological and petrophysical assessments, including porosity, permeability, and fluid distribution, are foundational to any CO_2 EOR initiative. These characteristics directly influence how CO_2 interacts with the reservoir, affecting both the efficiency of oil recovery and the stability of CO_2 storage. Projects that invested time and resources in reservoir characterization were better equipped to design and implement effective CO_2 injection strategies, leading to more successful outcomes.

Another significant takeaway is the optimization of injection strategies. Successful projects demonstrated that tailoring CO_2 injection methods to the specific conditions of each reservoir is essential for maximizing recovery efficiency. Techniques such as alternating CO_2 and water injection (WAG) or continuous CO_2 injection must be carefully selected and adjusted based on real-time data and reservoir behavior. By continuously optimizing these strategies, projects were able to improve oil recovery rates while minimizing CO_2 usage, thereby achieving both economic and environmental benefits.

The integration of advanced monitoring techniques also emerged as a best practice in successful CO_2 EOR projects. Real-time monitoring technologies, such as seismic surveys and production logging, were pivotal in optimizing CO_2 injection and managing production processes. These tools provided critical insights into CO_2 movement within the reservoir, allowing for timely adjustments and ensuring that the injection process remained efficient and effective. Advanced monitoring not only improved recovery outcomes but also enhanced the ability to safely store CO_2 over the long term.

Finally, the economic viability and risk management of CO_2 EOR projects were key factors in their success. Thorough economic assessments, including cost–benefit analyses and financial modeling, were essential for ensuring that projects were economically feasible. Additionally, effective risk management strategies, such as securing reliable CO_2 supplies and assessing market conditions, helped mitigate potential challenges. Projects that successfully balanced these economic considerations with technical execution were able to achieve favorable financial outcomes, making CO_2 EOR a viable and attractive option for enhancing oil recovery while contributing to carbon management goals.

6.2 Recommendations for future projects and research

Future CO_2 EOR projects should adopt a customized approach that considers the unique characteristics of each reservoir, utilizing advanced modeling and simulation tools to predict CO_2 behavior and optimize injection parameters. Given the increasing environmental concerns, it is crucial to focus on the sustainability of these projects by minimizing CO_2 emissions, enhancing the efficiency of CO_2 usage, and exploring synergies with CCS technologies. Additionally, enhanced collaboration and knowledge sharing among industry, academia, and government agencies are vital for the success of CO_2 EOR projects, as they facilitate the dissemination of best practices, technological advancements, and strategies for more efficient and economically viable operations.

7. Conclusions

CO_2 EOR plays within the broader framework of CCUS. It not only enhances oil recovery from mature fields but also serves as a viable method for sequestering large volumes of CO_2, thereby reducing greenhouse gas emissions. This dual benefit underscores the importance of CO_2 EOR as a crucial component in global efforts to mitigate climate change while addressing energy security.

Overall, the findings presented in this research work offer valuable insights into the practical and theoretical aspects of CO_2 EOR, providing a solid foundation for future research and development in this field. The ongoing advancement of CO_2 EOR technologies and practices holds significant potential for improving oil recovery, contributing to energy security, and playing a pivotal role in the global CCUS strategy for sustainable development.

Acknowledgements

The authors sincerely thank the UH Energy Industry partners for their support of the practical field research project.

Author details

Ganesh Thakur[1*], Xuejia Du[2] and Sabyasachi Prakash[2]

1 Energy Industry Partnership, USA

2 University of Houston, USA

*Address all correspondence to: gcthakur@uh.edu

References

[1] Faisal F, Tursoy T, Ercantan O. The relationship between energy consumption and economic growth: Evidence from non-granger causality test. Procedia Computer Science. 2017;**120**:671-675. DOI: 10.1016/j.procs.2017.11.294

[2] Moss T, Kincer J. How does energy impact economic growth? An overview of the evidence. Available from: https://Energyforgrowth.Org/Article/How-Does-Energy-Impact-Economic-Growth-an-Overview-of-the-Evidence/#:~:Text=The%20positive%20relationship%20between%20energy,Country%20that%20consumes%20a%20lot

[3] Nikitas G, Bhattacharya S, Vimalan N. Wind Energy. In: *Future Energy*. Elsevier; 2020. pp. 331-355. DOI: 10.1016/B978-0-08-102886-5.00016-5

[4] EIA EIA Projects 28% Increase in World Energy Use by 2040. Available online: https://www.eia.gov/energyexplained/us-energy-facts/ [Accessed: 6 August 2024]

[5] Lu G, Wang Z, Bhatti UH, Fan X. Recent Progress in carbon dioxide capture technologies: A review. *Clean energy*. Science and Technology. 2023;**1**:32-79. DOI: 10.18686/cest.v1i1.32

[6] Stewart C, Hessami M-A. A study of methods of carbon dioxide capture and sequestration—The sustainability of a photosynthetic bioreactor approach. Energy Conversion and Management. 2005;**46**:403-420. DOI: 10.1016/j.enconman.2004.03.009

[7] Marchetti C. On geoengineering and the CO2 problem. Climatic Change. 1977;**1**:59-68. DOI: 10.1007/BF00162777

[8] Feron PHM, Hendriks CA. CO2 capture process principles and costs. Oil and Gas Science and Technology. 2005;**60**:451-459. DOI: 10.2516/ogst:2005027

[9] Al-Mamoori A, Krishnamurthy A, Rownaghi AA, Rezaei F. Carbon capture and utilization update. Energy Technology. 2017;**5**:834-849. DOI: 10.1002/ente.201600747

[10] Li Y, Wang R, Zhao Q, Xue Z. Technological advancement and industrialization path of Sinopec in carbon capture, utilization and storage, China. *Energy Geoscience*. 2024;**5**:100107. DOI: 10.1016/j.engeos.2022.04.003

[11] Mantripragada HC, Zhai H, Rubin ES. Boundary dam or Petra Nova – Which is a better model for CCS energy supply? International Journal of Greenhouse Gas Control. 2019;**82**:59-68. DOI: 10.1016/j.ijggc.2019.01.004

[12] Tarufelli B, Snyder B, Dismukes D. The potential impact of the U.S. carbon capture and storage tax credit expansion on the economic feasibility of industrial carbon capture and storage. Energy Policy. 2021;**149**:112064. DOI: 10.1016/j.enpol.2020.112064

[13] Furre A-K, Meneguolo R, Ringrose P, Kassold S. Building confidence in CCS: From Sleipner to the northern lights project. First Break. 2019;**37**:81-87. DOI: 10.3997/1365-2397.n0038

[14] Trupp M, Frontczak J, Torkington J. The gorgon CO2 injection project – 2012 update. Energy Procedia. 2013;**37**:6237-6247. DOI: 10.1016/j.egypro.2013.06.552

[15] Tanaka Y, Sawada Y, Tanase D, Tanaka J, Shiomi S, Kasukawa T.

Tomakomai CCS demonstration project of Japan, CO2 injection in process. Energy Procedia. 2017;**114**:5836-5846. DOI: 10.1016/j.egypro.2017.03.1721

[16] Sheikh F. Commercialization of Al Reyadah – World's 1st carbon capture CCUS project from iron and Steel industry for enhanced oil recovery CO_2-EOR. In: Proceedings of the Day 2 Tue, November 16, 2021. Abu Dhabi: SPE; 9 Dec 2021

[17] Denduluri, V.; Thakur, G. Carbon Capture and CO_2 EOR/Storage—A Game Changer CCUS Technology. Available from: Https://Jpt.Spe.Org/Twa/Carbon-Capture-and-Co2-Eor-Storage-a-Game-Changer-Ccus-Technology

[18] Cheraghi Y, Kord S, Mashayekhizadeh V. Application of machine learning techniques for selecting the Most suitable enhanced oil recovery method; challenges and opportunities. Journal of Petroleum Science and Engineering. 2021;**205**:108761. DOI: 10.1016/j.petrol.2021.108761

[19] Yongmao H, Zenggui W, Binshan J, Yueming C, Xiangjie L, Petro X. Laboratory investigation of CO2 flooding. In: Proceedings of the Nigeria Annual International Conference and Exhibition. Abuja, Nigeria: SPE; 2004

[20] Gozalpour F, Ren SR, Tohidi B. CO2 EOR and storage in oil reservoir. Oil and Gas Science and Technology. 2005;**60**:537-546. DOI: 10.2516/ogst:2005036

[21] Li L, Zhao N, Wei W, Sun Y. A review of research Progress on CO2 capture, storage, and utilization in Chinese Academy of Sciences. Fuel. 2013;**108**:112-130. DOI: 10.1016/j.fuel.2011.08.022

[22] Preston C, Whittaker S, Rostron B, Chalaturnyk R, White D, Hawkes C, et al. IEA GHG Weyburn-Midale CO2 monitoring and storage project–moving forward with the final phase. Energy Procedia. 2009;**1**:1743-1750. DOI: 10.1016/j.egypro.2009.01.228

[23] Ampomah W, Balch RS, Grigg RB, McPherson B, Will RA, Lee S, et al. Co-optimization of CO2-EOR and storage processes in mature oil reservoirs. Greenhouse Gases: Science and Technology. 2017;**7**:128-142. DOI: 10.1002/ghg.1618

[24] You J, Ampomah W, Sun Q, Kutsienyo EJ, Balch RS, Dai Z, et al. Machine learning based Co-optimization of carbon dioxide sequestration and oil recovery in CO2-EOR project. Journal of Cleaner Production. 2020;**260**:120866. DOI: 10.1016/j.jclepro.2020.120866

[25] Kalteyer J. A case study of SACROC CO2 flooding in marginal pay regions: Improving asset performance. In: Proceedings of the Day 1 Mon, August 31, 2020. Virtual: SPE; 30 Aug 2020

[26] Thakur GC. Waterflood surveillance techniques - a reservoir management approach. Journal of Petroleum Technology. 1991;**43**:1180-1188. DOI: 10.2118/23471-PA

[27] Thakur GC. The role of reservoir Management in Carbonate Waterfloods. In: Proceedings of the all Days. New Delhi, India: SPE; 17 Feb 1998

[28] Satter A, Thakur G.C. Integrated Petroleum Reservoir Management: A Team Approach. Tulsa: PennWell; 1994

[29] Du X, Thakur GC. Lessons learned from the process of water injection Management in Impactful Onshore and Offshore Carbonate Reservoirs. Energies (Basel). 2024;**17**:3951. DOI: 10.3390/en17163951

[30] Zaluski W, El-Kaseeh G, Lee S-Y, Piercey M, Duguid A. Monitoring technology ranking methodology for CO2-EOR sites using the Weyburn-Midale field as a case study. International Journal of Greenhouse Gas Control. 2016;**54**:466-478. DOI: 10.1016/j.ijggc.2016.06.012

[31] Yuan S, Ma D, Li J, Zhou T, Ji Z, Han H. Progress and prospects of carbon dioxide capture, EOR-utilization and storage industrialization. Petroleum Exploration and Development. 2022;**49**:955-962. DOI: 10.1016/S1876-3804(22)60324-0

[32] Advanced Resources International The U.S. CO2 Enhanced Oil Recovery Survey an Interim Update of Enhanced Oil Production Totals and CO2 Supplies for Active CO2 EOR Projects in the U.S. as of End-of-Year 2022. Available from: https://Adv-Res.Com/Pdf/2022-CO2-EOR-Survey-Brochure-FINAL-FEB-26-2024.Pdf

[33] Du X, Salasakar S, Thakur G. A comprehensive summary of the application of machine learning techniques for CO2-enhanced oil recovery projects. Machine Learning and Knowledge Extraction. 2024;**6**:917-943. DOI: 10.3390/make6020043

[34] Chen P, Bose S, Selveindran A, Thakur G. Application of CCUS in India: Designing a CO2 EOR and storage pilot in a mature field. International Journal of Greenhouse Gas Control. 2023;**124**:103858. DOI: 10.1016/j.ijggc.2023.103858

[35] Chen P, Selveindran A, Kumar C, Saloma Y, Bose S, Balasubramanian S, et al. CO2-EOR and carbon storage in Indian oilfields: From laboratory study to pilot design. In: Proceedings of the Day 2 Wed, April 24, 2019. San Jose, California, USA: SPE; 22 Apr 2019

Chapter 5

Assessing Baseline Groundwater Quality near a Proposed Direct Air Capture (DAC) Facility

Vinay Kumar and Vihaan Rajput

Abstract

This study evaluated baseline groundwater quality near the proposed Direct Air Capture (DAC) facility in Kingsville, TX, USA. It hypothesized that groundwater quality in the region has remained relatively stable over time. Rainwater and groundwater data from NOAA, along with records from approximately 30 water wells, were analyzed. A detailed assessment was conducted on Well ID 8329702, which had the most extensive and continuous dataset for alkalinity, bicarbonate, pH, and hardness from 1973 to 2012. The analysis revealed a slight decline in alkalinity (−0.00006) and bicarbonate (−0.007) over time, while changes in pH and hardness were minimal. These findings provide a critical baseline for future groundwater monitoring, particularly in areas surrounding industrial facilities such as DAC plants.

Keywords: direct air capture, ground water wells, alkalinity, hardness, bicarbonate

1. Introduction

The IPCC has suggested reducing human-induced CO_2 emissions to near-zero to address global warming [1]. Furthermore, efforts to limit global warming to 1.5°C, as targeted in the 2015 Paris Agreement, are estimated to have the potential to capture between 350 and 1,200 gigatons of CO_2 by 2100 [2]. The United Nations' Sustainable Development Goals (SDGs) outline 17 objectives to achieve global sustainability. Carbon capture and storage/utilization (CCS/CCU) directly aligns with goals such as Climate Action (Goal 13) and fostering innovation & industrial growth (Goal 9). Technologies for reducing CO_2 emissions are critical in the fight against climate change. These approaches range from capturing emissions at the source to removing CO_2 already in the atmosphere. These efforts aim to stabilize or reduce the rising global air temperature on Earth. Over the past ~135 years, there has been an increase in CO_2 levels, accompanied by a corresponding rise in air temperature (**Figure 1**).

Naturally respiration from flora and Funa, volcanic eruptions, wildfires, decomposition of organic matters, and weathering are the main sources of carbon in the earth system. These are the beginning of carbon's journey. On the other hand, ocean, plants (*via* photosynthesis), soil, carbonate rocks, and wetlands-peatlands are the main sink of carbon. Plants play an important role in capturing carbon from the atmosphere and transforming carbon into organic matter.

IntechOpen

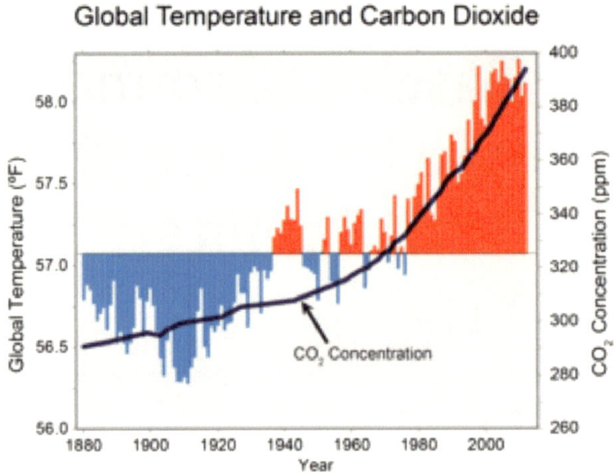

Figure 1.
Global variation of temperature and CO_2 (Solid black line) from 1880 to 2012. Blue and red bars show temperatures below and above long-term average (Source: Ref. Karl et al. [3]).

In earth system, carbon cycle is being balanced somehow. However, industrialization and transportation due to demand of materialistic stuff have brought an imbalance in this cycle of carbon.

One of the technology, carbon capture and storage (CCS) captures CO_2 from point sources such as power plants or industrial facilities and stores it in geological formations. It is particularly relevant for hard-to-abate industries like steel and cement productions that account for a significant share of global emissions [4]. Another one carbon capture, utilization, and storage (CCUS), which is beyond storage, this approach involves using captured CO_2 to create valuable products like chemicals, fuels, or construction materials, helping offset the costs of capture while promoting circular economy practices [4].

Some other methods and technologies are natural, for example, biological carbon sequestration, which includes strategies like afforestation, reforestation, and soil carbon enhancement to naturally absorb CO_2 through ecosystems (**Figure 2**). These methods are often aligned with sustainable land use practices [6]. Next one in the same row is enhanced weathering involves spreading reactive minerals over land to absorb CO_2 from the atmosphere through chemical reactions, effectively accelerating natural geological processes.

One of the most prominent technologies, for example, direct air capture (DAC) technology, uses engineered systems to extract CO_2 directly from the ambient air. Once captured, the CO_2 can either be stored underground to achieve negative emissions or utilized in industrial processes, such as synthetic fuel production. DAC systems have seen significant advancements in the recent years and hold promises for scaling up to combat dispersed emissions effectively [7, 8]. Despite its potential, widespread deployment faces economic and regulatory challenges.

Each of these technologies plays a unique role in achieving net-zero emissions and mitigating global warming. They are supported by international frameworks such as the United Nations' Sustainable Development Goals and the Paris Agreement, which emphasize their importance in reaching climate targets [9].

Groundwater quality is a crucial factor in maintaining the health and sustainability of ecosystems and human activities, particularly in regions undergoing industrial

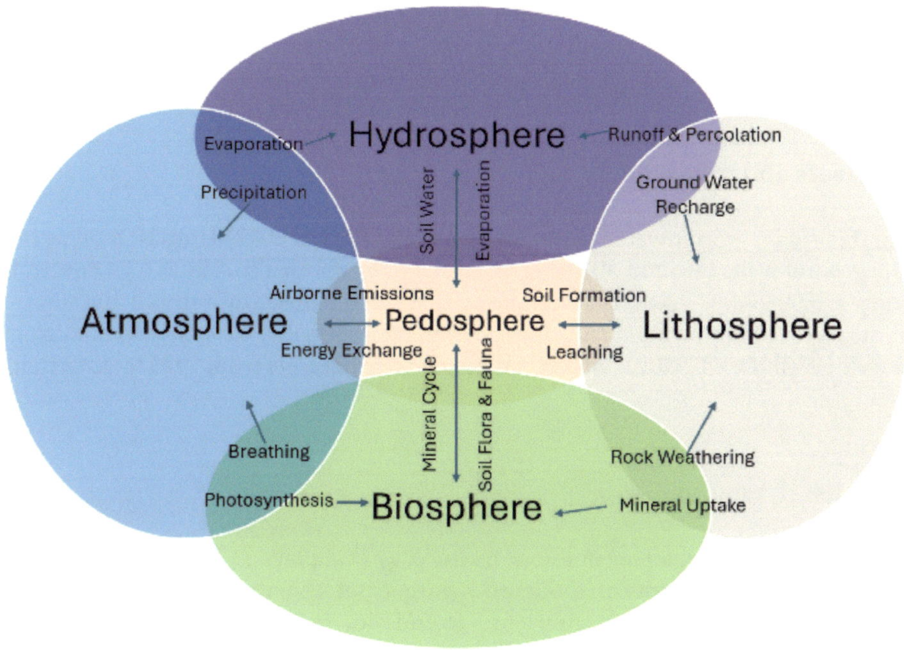

Figure 2.
A complex but synergic interaction between the various components of Earth system (Inspired by Lal et al. [5]).

development. The introduction of industrial facilities, such as direct air capture (DAC) plants, necessitates a thorough understanding of the existing environmental conditions, especially groundwater quality, to mitigate potential adverse impacts. DAC technology, which is aimed at capturing carbon dioxide directly from the atmosphere, represents a significant step forward in combating climate change; however, the establishment of such facilities also raises concerns about their potential impact on surrounding water resources.

The Kingsville area in south Texas, identified as a proposed site for a DAC facility, presents a unique opportunity to assess the baseline groundwater quality before the plant becomes operational. Understanding the historical trends in water quality parameters such as alkalinity, bicarbonate levels, pH values, and hardness is essential for monitoring changes that may occur post-establishment of the DAC facility. Prior studies have shown that industrial activities can influence local groundwater by altering its chemical composition, potentially leading to increased acidity, hardness, or the presence of contaminants [10].

By documenting the historical groundwater quality trends, this study aims to contribute valuable insights into the potential impacts of DAC plants on surrounding environments, emphasizing the need for continuous and comprehensive environmental monitoring as industrial activities expand [11]. The findings of this research underscore the importance of proactive environmental management in regions undergoing significant industrialization.

Groundwater and rainwater data for many decades to understand the trends might be vital for environmental protection but also for ensuring the long-term viability of water resources in the region. The hypothesis in general remains that the baseline water quality has remained mostly constant over time in the study area. However, the proposed direct air capture (DAC) region's groundwater quality trends and the factors that may influence them require further investigation. With the addition of the plant being established near the Kingsville area, the potential impacts around it must be monitored.

2. Datasets and methods

In this study, groundwater and rainwater data (1973 to 2012) from NOAA and many ground wells (around 30) were analyzed to establish a baseline of groundwater quality in Kingsville, Texas, USA. Additional gridded datasets of rainfall and temperature were obtained from NOAA for comparison. Finally, we analyzed data from one well (Well ID: 8329702), which had continuous datasets from 1973 to 2012 than others.

3. Results

DAC is a critical negative emissions technology that plays a vital role in achieving the 1.5–2°C global temperature moderation targets set under the Paris Agreement [12]. DAC serves as a complementary tool to emission reduction strategies by actively removing carbon dioxide (CO_2) from the atmosphere, addressing residual emissions that are difficult to eliminate through decarbonization alone. **Figure 3a** shows the locations of approximately 25 underground water wells (blue dots) surrounding the DAC site (green diamond), with well #8329702 highlighted by a red star. We studied in this chapter is marked with red star over south Texas, USA. The location of DAC has moderate pH as compared to other locations in **Figure 3b**. Such investigation on the pH of water well will provide as the baseline, if due to construction of DAC will affect the pH of water well in future. Such change in pH might be due to leakage of the stored CO_2 in the DAC facility. The leakage of carbon from DAC facility might be due to numerous things which we are not trying to discuss here.

The water quality in South Texas is significantly influenced by its underlying geology, particularly the presence of limestone formations. As a result, water in this region typically exhibits moderate-to-high alkalinity, which stabilizes pH and buffers against acidification. The bicarbonate content is also elevated due to the dissolution of carbonate rocks, contributing to the water's hardness and pH regulation. pH values generally range from 6.5 to 8.5, as established by Texas Surface Water Quality Standards, ensuring the protection of aquatic life and human health. Water hardness, driven by the presence of calcium and magnesium ions, is classified as "moderately hard" to "hard," affecting household water usage but not posing health risks. These line diagrams show well number 8329702's variability over 1973–2012 (**Figure 4**). There is a small decrease over time of by the rate of −0.00006, proving that the alkalinity in this well decreased over time (**Figure 4a**). Next scatterplot shows the decrease rate of bicarbonate by −0.007 over the study time (**Figure 4b**). In **Figure 4c**, scatterplot shows pH values of the well's water over 1973–2012. There is a very small change in the value over this period. Hardness decreased at a rate of −0.0004 per year since 1973.

However, in this study, only four key parameters were chosen based on their direct relevance to the specific objectives of the research and the local groundwater conditions. These parameters were selected to provide a clear and focused analysis of the most critical factors affecting groundwater quality in the study area. While other parameters like pH, turbidity, and heavy metals are important for a comprehensive

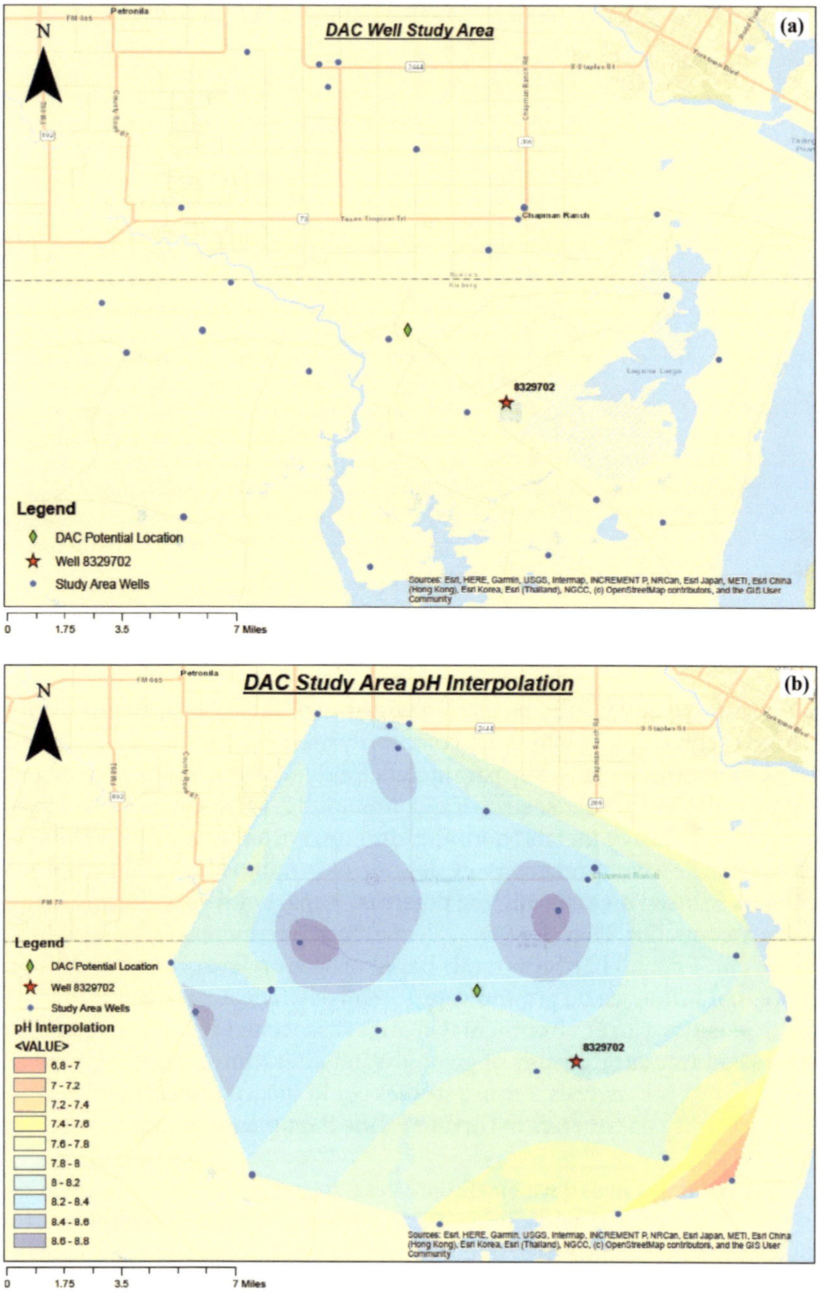

Figure 3.
(a) DAC potential location (green diamond), well 8,329,702 (red star), and water wells (blue dots) around, (b) Ph interpolation around DAC study area.

assessment, the chosen four were prioritized due to their significant impact on the study's outcomes and the availability of reliable data. This approach ensures a targeted and manageable analysis while still addressing the core concerns of groundwater quality in the context of this research.

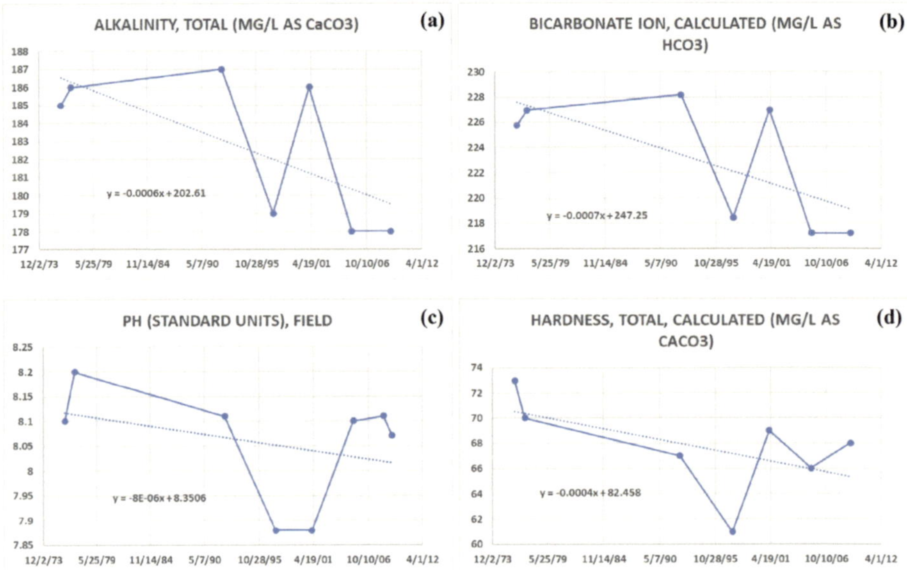

Figure 4.
(a) Alkalinity total, (b) bicarbonate Ion, (c) Ph, and (d) hardness total for Well ID: 8329702 near DAC.

Groundwater quality is influenced by a wide range of physical, chemical, and biological parameters, many of which are commonly used in drinking water assessments. As noted by Dohare et al. [13], key parameters for drinking water quality include pH, turbidity, total dissolved solids, electrical conductivity, total hardness, and various dissolved ions, and heavy metals (boron, cadmium, chromium...etc.). While these parameters provide a comprehensive evaluation of groundwater quality, the scope of this study was focused on assessing the potential impact of direct air capture (DAC) on groundwater quality. Therefore, we selected four parameters (alkalinity total, bicarbonate ion, Ph, and hardness total) based on their relevance to the DAC process and its potential influence on groundwater chemistry. These parameters were chosen due to their sensitivity to environmental changes associated with DAC operations, their established role as indicators of groundwater alterations, and the availability of reliable datasets for analysis. Future studies could incorporate a broader range of groundwater quality parameters to further refine the understanding of DAC-related impacts.

South Texas experiences a warm climate year-round, with average annual temperatures ranging from the mid-60s°F (18°C) in winter to the mid-90s°F (35°C) in summer. Rainfall is unevenly distributed, with the wettest periods occurring in late spring, early fall, and during tropical storm activity, leading to annual precipitation totals of 20 to 40 inches. The region faces high rainfall variability, often experiencing prolonged droughts that challenge water resource management and agriculture. Climate change is expected to exacerbate this variability, increasing the frequency of extreme weather events such as heavy downpours and prolonged dry spells, thereby affecting water availability and ecosystem health. **Figure 5(a,b)** shows the spatial distribution of rainfall and surface temperature. Long-term variations in the rainfall and temperature can influence the total water availability and characteristics of water in ground wells around the DAC facility and south Texas region.

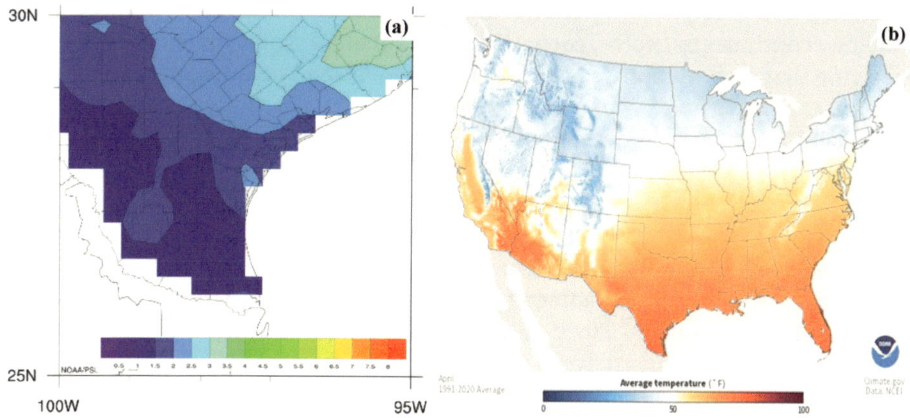

Figure 5.
(a) April climatology (1991–2020) of rainfall (NOAA-CPC Unified Gauge-based) over the coastal bend region of Texas and (b) April temperature (1991–2020) over the United States of America.

4. Discussion

The findings of this study provide important insights into the baseline groundwater quality in the Kingsville, Texas region, particularly in the context of the proposed Direct Air Capture (DAC) facility. The analysis of alkalinity, bicarbonate, pH, and hardness over a 40-year period (1973–2012) reveals a gradual decline in alkalinity and bicarbonate levels, alongside minimal changes in pH and hardness. These trends suggest that groundwater in the region is becoming slightly more acidic over time, which could have implications for both environmental and industrial activities. While the changes are small, they highlight the need for continuous monitoring, especially as the DAC facility becomes operational and potentially interacts with local groundwater systems.

The observed decline in alkalinity and bicarbonate may be attributed to natural processes such as the dissolution of carbonate minerals or anthropogenic influences, including agricultural runoff or changes in land use. The stability of pH and hardness, however, indicates that the groundwater system remains relatively buffered against significant acidification. This is consistent with the region's geology, which is dominated by limestone formations that naturally regulate pH and hardness. Nevertheless, the introduction of a DAC facility could alter these dynamics, particularly if CO_2 leakage occurs during storage operations. Even minor changes in groundwater chemistry could affect local ecosystems, water availability, and the long-term viability of the DAC facility itself. Therefore, establishing a robust monitoring framework that includes a broader range of parameters (e.g., heavy metals, turbidity, and dissolved gases) will be essential for detecting and mitigating potential impacts.

This study underscores the importance of baseline groundwater quality assessments in regions undergoing industrial development. While the findings are based on data from a single well, they provide a foundation for future research and monitoring efforts. Expanding the dataset to include multiple wells and additional parameters would improve the generalizability of the results and offer a more comprehensive understanding of groundwater trends in the region. Furthermore, integrating climate change projections into future analyses could help anticipate how shifts in temperature and precipitation patterns might influence groundwater quality. As DAC technology continues to evolve, proactive environmental management will be critical to ensuring that carbon capture efforts do not inadvertently compromise local water resources.

5. Conclusion

In this study, groundwater and rainwater data from NOAA and many ground well were analyzed to establish a baseline of groundwater quality in Kingsville, Texas, USA. The data span from 1973 to 2012 and include key parameters that can indicate shifts in water quality. This analysis provides a foundation for future monitoring efforts, enabling the detection of any deviations from the established baseline that could be attributed to the operation of the DAC facility. Understanding these trends is vital not only for environmental protection but also for ensuring the long-term viability of water resources in the region. Based on the current data trends and sets presented, the overall values for all four variables are decreasing gradually over time. This shows that the water quality is progressively becoming more acidic, which can be affected by many factors. The well-studied, 8,329,702, showed many different trends across different variables. These trends were also shown in other wells across the region. The hypothesis that groundwater quality has remained relatively stable was supported by the data. For future research on the topic, we would like to look at more specific parameters like certain wells or areas and we would also like to look at certain chemicals that can affect the water quality in those wells.

Acknowledgements

The author acknowledges the use of ChatGPT for language polishing.

Author details

Vinay Kumar[1*] and Vihaan Rajput[2]

1 Department of Atmospheric Science, Environmental Science and Physics, University of the Incarnate Word, San Antonio, TX, USA

2 V.M.H.S. Corpus Christi, TX, USA

*Address all correspondence to: vkumar@uiwtx.edu

References

[1] IPCC Report. Climate Change 2022 - Mitigation of Climate Change: Working Group III Contribution to the Sixth Assessment Report of the Intergovernmental Panel on Climate Change. 1st ed. Cambridge, UK: Cambridge University Press; 2023. DOI: 10.1017/9781009157926

[2] IPCC Report. Global Warming of 1.5°C: IPCC Special Report on Impacts of Global Warming of 1.5°C above Pre-Industrial Levels in Context of Strengthening Response to Climate Change, Sustainable Development, and Efforts to Eradicate Poverty. 1st ed. Cambridge, UK: Cambridge University Press; 2022. DOI: 10.1017/9781009157940

[3] Karl TR, Melillo JT, Peterson TC. Global Climate Change Impacts in the United States. New York, NY: Cambridge University Press; 2009. p. 189

[4] Bui M, Adjiman CS, Bardow A, Anthony EJ, Boston A, Brown S, et al. Carbon capture and storage (CCS): The way forward. Energy & Environmental Science. 2018;**11**(5):1062-1176

[5] Lal R et al. Pedospheric processes and the carbon cycle. In: Lal R et al., editors. iSoil Processes and the Carbon Cycle. Boca Raton, FL: CRC Press; 1998. pp. 1-8

[6] Mahboob S, Govindarajan M, Al-Ghanim KA, Ahmed Z, Al-Mulhm N, Vodovnik M, et al. A review on biological carbon sequestration: A sustainable solution for a cleaner air environment, less pollution and lower health risks. Saudi Journal of Biological Sciences. 2021;**28**(3):1801-1810

[7] Beuttler C, Charles L, Wurzbacher J. The role of direct air capture in mitigation of anthropogenic greenhouse gas emissions. Frontiers in Climate. 2019;**1**:10

[8] McQueen N, Psarras P, Pilorgé H, Liguori S, He J, Yuan M, et al. Cost analysis of direct air capture and sequestration coupled to low-carbon thermal energy in the United States. Environmental Science & Technology. 2021;**55**(8):5135-5144

[9] Morris J, Chen YH, Gurgel A, Reilly J, Sokolov A. Net zero emissions of greenhouse gases by 2050: Achievable and affordable. Journal of Environmental Economics and Management. 2023;**118**:102767

[10] Basu A, Saha D, Saha R, Ghosh T, Saha B. A review on sources, toxicity and remediation technologies for removing arsenic from drinking water. Research on Chemical Intermediates. 2014;**40**(2):447-485

[11] Madhu K, Pauliuk S, Dhathri S, Creutzig F. Understanding environmental trade-offs and resource demand of direct air capture technologies through comparative life-cycle assessment. Nature Communications. 2021;**12**(1):3965

[12] Minx JC, Lamb WF, Callaghan MW, Fuss S, Hilaire J, Creutzig F, et al. Negative emissions—Part 1: Research landscape and synthesis. Environmental Research Letters. 2018;**13**(6):063001

[13] Dohare D, Deshpande S, Kotilya A. Analysis of ground water quality parameters: A review. Research Journal of Engineering Sciences. 2014;**3**(5):26-31